Acknowledgments

The author would like to thank:

Friends and relatives who helped in various ways with recent (since 1998) Hartland books: Roger C.; my parents; brother John; collector Cecile Bellmer, for a critical reading; and artist Michael Mueller, who helped with photo selection.

Schiffer Publishing Ltd.

The people of Hartland Plastics, 1939-1978, especially: Paul & Bernice Champion, Alvar & Henny Bäckstrand, Edwin & Josephine Hulbert, Robert & Rita McGuire, Roger & Idella Williams, Thomas Caestecker, Hans Seuthe, Jerome Delsman, Richard Petfalski, Janet Gerbenskey, and Richard Rohde (son of Armin Rohde).

Personnel, past or present, of more recent Hartland companies, especially: Mr. Bev W. Taylor, Mark Borzillo, Liz Dalpe, Kim Lorraine, Russell Seifert, Tina Strubberg, Don Light, and Patty Heldt of Steven Mfg.; Paola Groeber; sculptors Carol Gasper, Linda Lima, and Kathleen Moody/Elite Decorations; and officials at Strombecker Corp. My thanks goes to all the past or present owners and employees of Hartland companies (or their relatives) who shared their story and, in some cases, provided or permitted photographs.

Terry Biwar Becker, Waukesha County (Wis.) Museum Librarian, Pam Weinhammer, Hartland (Wis.) Historian, and Nancy Massnick, Hartland Public Library; Mary Dulebohn, Hermann Branch of the Scenic Regional Library; Erin Renn, Deutschheim State Historical Site, Hermann, Missouri; Don Kruse, editor of the *Hermann Advertiser-Courier;* Carolyn Laboube, H.O.M.E., Inc., Hermann; Dan Reynolds, Anheuser-Busch, St. Louis, Missouri, and officials at Nylint Corp., Rockford, Illinois.

Moden Plastics magazine, a publication of Chemical Week Associates, Inc., for permission to reprint a photo and brief portions of text from the October 1960 issue.

Many attentive photo developers/printers, especially Diane, Chris R., Daisie, and Eric at Walgreens' on Brady St.; and Helix Photoart for photography advice.

These collectors who contributed photographs or made it possible for their models to be photographed for this book (including five people who shipped models to me): Donna Anderson, Beth Andries, Sandy Bellavia, Cecile Bellmer, Vicky Boehm, Jaci Bowman (The Silver Leash), Carla S. Clifford, Terry Davis, Laura Diederich, Lea Dobranski, Traci Durrell-Khalife, Kelly (Engelsiepen) Scotti, Tina English-Wendt, Kim Fairbrother, Eleanor & Shay Goosens, Michelle Grant, Judy & Kelley Harding, Eleanor Harvey, Jackie Himes, Peggy Howard, Maggi Jacques, Stephanie Jones, Shirley Ketchuck, Jan Kreischer, Daphne R. Macpherson (Cascade Models), Charlene Marshall, Barri Mayse, Heather M. McCurdy, Elizabeth McMaster, Judith Miller, Bobbie Mosimann, Michael Mueller, Ingrid Muensterer (Modell Pferde Versand), Vickie Neiduski, Karen Oelkers, Laura Pervier (Lone Wolf Star), Judy Renee Pope, April and Jon Powell, Sande Schneider, Michelle Smalling, Anni Stapley-Koziol, Jacqueline Tierney, Sandy Tomezik, James W. and Sandra J. Truitt, Bonnie Valentine (Breyers by Mail), Ellen W. Vogel, Suzanne Wadke-Orton, Laura K. Whitney, Pam Young, and Susan Bensema Young.

These additional collectors, for information (including participation in *Hartland Market* surveys) or encouragement: Cheryl Abelson, Chris Anderson, Joyce Anderson, Dana Bennett, Jeannine Bergeron, Mark Blackwell, Brian Blauch, Elaine Boardway, Bettye Brown, Debbie Buckler, Sheila Callahan, Katie Carter, Denise Chance-Hauck, Peggy Collins, Karen Crossley, Janet Davis, Bonnie Ellis, Jim Foley, Debbie Gamble-Arsenault, Cheryl Greene, Karen Grimm (Black Horse Ranch), Andrea Gurdon, Lydia (Gutierrez) Fey, Laurie Jo Jensen, Kim Lory Jones, Kim Kadlec, Jeannie (Thomas) Kelly, Nancy Kelly, Diana Krafik-Montgomery, Sue Lehman, Daria Littlejohn, Gay Mahlandt, Jo Maness, Janet Marshall, Betty Mertes, Brenda Metcalf, Alisha Miller, Debbie Moore, Torry Morgan, Lindy Pinkham, Jenny Provencher, Gayle Roller, Sue Rowe, Carole & Julia Schwartz, Kitty Steen, Marte Stines, Liz Strauss, Shari Struzan, Lillian Sutphin, Connie Thron, Rick Van Etten, Marney Walerius, Linda Walter, Heather Wells, Christine & Michelle Wilder (The Happy Cayuse), and Nancy A. Young.

Photo Credits. Where no credit is given with a model's photo, it means that the author was both the photographer and the owner of the model. If the photo credit reads, "Model, courtesy of...," it means that the person whose name is mentioned owned the model and made it available for the author to photograph. If a photo credit reads, "Courtesy of...," it means that the person named was both the owner and photographer.

Contents

I never intended to collect model horses, but that's where my interests in horses and art intersected. In the 1960s, I admired model horses in stores, catalogs, and at friends' houses, but spent most of my allowance on riding, instead. In the 1970s, I wished I had bought more model horses, and started to look for them in stores. When I called Strombecker Corporation (the second Hartland company) in 1974, I found out that Hartland production had ended the year before.

In 1975, I joined the community of model horse collectors created by Linda Walter's *The Model Horse Shower's Journal* (and Breyer's *Just About Horses),* and bought model horses through its pages. In 1980, Cheryl Abelson's *Hagen-Renaker Handbook* broke new ground as the first model horse book and showed how self-publishing could be done.

I liked and owned not just Hartlands, but examples of all the major (and many minor) brands of models horses. Color catalogs for Breyers, Beswicks, and others were available, but information on Hartland models was very limited. Some collectors sold black-and-white photocopies of a few, old Hartland catalog sheets without dates, and "articles" in model horse journals were—except for two articles by Sandy Tomezik—usually only a few sentences long. After made-in-Japan china horses, the greatest need for information was on Hartlands, and they had been made only a county away from my home. I wanted to answer my own questions about Hartland, knowing that others would be interested in the answers, too. So, I started the research in fall of 1980, and first made contact with former Hartland personnel, whom I had located from scratch, in January of 1981. I had just finished several years of work in magazine publishing, and in research and writing in local government, so I felt qualified to undertake a self-published book.

The first edition, with 26 pages of color photos (72 pages in all), came out in February 1983. I did annual printings for six years, and printed two or three times per year from 1991-1995. The book was advertised in model horse collecting publications, but was also purchased by gentlemen who collected the rider sets and told their friends. More than half of the hundreds of photos were always printed in color. I published often, in small quantities, because I could not afford to print 500 or 1,000 books at a time. Because of the color and the small press runs, it was a very expensive book (per unit). However, I sold it on a nonprofit basis, my personal resources were continuously tied up with it, and I gave it priority over my own collection. Fortunately, other collectors have always sent photos of models I didn't own or have made their models available for me to photograph. (I love photographing model horses.)

This book has been a full-time effort, on and off, over the last 20 years. The first edition took a solid year to write, design, and produce. Revisions for the first five editions after that took a total of 20 months, working at it full-time. Each printing took an additional week. The 1998 book was eight months in the making. *Hartland Horsemen* (1999) and this book, together, occupied my time for almost two years. My equipment has evolved from a manual typewriter (a 1936 Royal) to advanced, desktop publishing software (for the 1998 edition) that still has me outsmarted. All books earlier than 1999 were hand-collated by me. Different portions of the book were not only printed on different machines, but by different printing companies because I tested and selected the best machines for color, black-and-white text (including pasted-up pages), and photo reproductions in black-and-white. One printing could involve so many as fourteen trips to three different printers. Usually, this meant I was hauling heavy boxes of printed pages (or bound books) back and forth on the bus although I could walk to the printer across the street.

I guess you could say that the love of horses—along with the help of friends and the determination to accomplish something worthwhile—can work wonders.

Based on Original Research

This book is based on original research. Since 1980, I've interviewed more than a dozen people from the original Hartland company, several from Steven Mfg., and people from each of the two other Hartland horse companies (and two other Hartland baseball statue companies). With all of the information I've gathered—especially since 1989—from in-person interviews, phone interviews, correspondence, and even an oral history (tape) from owners, executives, sculptors, and other personnel of Hartland Plastics and later Hartland companies, I feel that I have an accurate idea of the Hartland story.

In addition, I've studied paper resources: company catalogs and advertisements, vintage newspaper articles on the Hartland companies, and public (government) records. I've done "field research" at flea markets and antique stores, have conversed and corresponded with other collectors, bought and sold models through publications, and since March of 1998, have looked at Internet auctions daily, where 200+ Hartland items are sold every day.

This book has been sold to collectors and model horse manufacturers in the U.S. and abroad since 1983, and has had the benefit of about 17 years of feedback from reader-collectors. Reconstructing the history of a defunct company is never easy, but my account of the original Hartland company's history has been excellent since 1990.

Newsletter and Other Articles. From December 1994—May 1996, I published 18 monthly issues of *Hartland Market*, a newsletter for collectors. It totaled 194 pages, including the indexes, and most of it was editorial material, rather than advertising. I wrote news and history updates from my ongoing research, and readers contributed articles and letters and answered surveys whose results I compiled, interpreted, and reported. Results of my survey on Tennessee Walkers are referenced in this book. I also have written articles on Hartland models for model horse journals and toy collecting magazines, and I was a consultant to Steven Manufacturing. By special arrangement, some quotes from my book appeared on the 1993-1994 Hartland/Steven boxes and catalogs.

The Seminal Source. Similarities between this book and material on Hartland written by others are seldom a coincidence. This book has been available for a long time, and for about 14 years, it was the only information in print about Hartland that was longer than a couple of pages. Just about all writers on Hartland model horses since 1983 have purchased this book first. Some of them failed to acknowledge sources or to distinguish their writing from mine. The same is true of some materials, including recent web sites, read by Hartland western or sports collectors. In one instance, an individual sold collectors material exactly copied from my book, and a toy magazine published it without knowing its true origin. All of my self-published editions and other writings are copyright-registered, and no one had any business copying them.

The Truth About Hartland Plastics

Some Hartland materials published by others include painfully incorrect "facts" obtained from a former employee of Hartland Plastics who, for 20 years, has posed as an artist/designer and product concept originator, but wasn't. I should know! Before I made contact with the owners and executives of Hartland Plastics and the company's two sculptors—Roger Williams and Alvar Bäckstrand, he had me fooled, too. Early editions of my book were tainted with not only his false claims of importance, but also with his "facts" about models that, later, proved untrue.

In addition, his incorrect information threw me off the trail of locating Hartland sculptor Roger Williams. He told me in 1981 that Roger Williams had passed away, and so I didn't look for him. Later, the impostor told the same story to a sports writer whose article appeared in *The Milwaukee Journal,* June 25, 1987. The article says, "[He] not only dreamed up the idea of producing the statues, he designed them." Later in the article, he's quoted as saying,"...mostly I made drawings just by looking at the players. Roger Williams of our company did the sculpting. He died many years ago." The newspaper then heard from Mrs. Williams, who informed them that her husband of 40 years was alive and well, and it printed a correction on June 27, 1987. They were living in the *Milwaukee Journal* readership area, only a few miles from the impostor. It took three more years before I connected with Mr. and Mrs. Williams. Then, they laughed at the absurd claims in the newspaper article, but I didn't think it was funny.

In 1996, Alvar Bäckstrand's comment was, "Only Roger Williams and I were the sculptors at Hartland Plastics, Inc. Anyone who tells anything else or pretends to have had anything to do with the design and making of the original models, in clay and model metal, is ...."

Since 1990, my books (and other writings) have set the facts straight, but, ten years later, I still read the same misinformation about Hartland, quoting the same individual, in many places. The person who dreamed up the idea of making Hartland baseball statues was in fact Thomas E. Caestecker, who with his parents, owned Hartland Plastics. The designers of the baseball statues were in fact the sculptors, Alvar Bäckstrand and Roger Williams. My 1995 Hartland newsletter (and 1991-1995 books) published Alvar Bäckstrand's detailed account of showing a clay model-in-progress to a baseball player. A collector said in 1996 that a person other than Mr. Bäckstrand had just told him the story, pretending to be the sculptor. The imposter had received copies of my 1983 and 1991 books, and the newsletter.

The truth about Hartland Plastics is what it should be: the artists, owners, and executives of Hartland Plastics were—although it may be unfashionable to say this—people of character: honest, talented, and hard-working. They don't seek attention for what they did, much less for what they didn't do. The people who seek adulation are seldom the ones who have earned it.

Wish List. I wish that: (1) this book included a photo of small-scale manufacturer Paola Groeber (who was born in the early 1950s); (2) Paola's job had allowed her more time to compare my model lists/descriptions with her Hartland records (a year and a half wasn't enough time, but I'm glad she was able to answer some questions); (3) employees of Steven Manufacturing Company (the third Hartland manufacturer), could have been able to supply information about the quantities of models they made (but some computer disks couldn't be found, others would be difficult to access due to a change in computer systems, and the current owner reportedly forbade them to help with this book).

Fortunately for the story of Hartland, the previous owners and employees of Steven Manufacturing Company and various residents of Hermann, Missouri, were very helpful in contributing to this book, and I admire their spirit. I'm also grateful to Strombecker Corp. (the second Hartland company) and to the many owners and employees of the original Hartland Plastics company who contributed so much.

Because of my long-term research of Hartland models and the companies that made them, I now count some former owners and employees of Hartland as my friends. I wish you could know what fine people they are. This book is a celebration of them, of horses, of good ideas, and of fine craftsmanship, and none of those things should be taken for granted.

Introduction to Hartland Models

From the mid-1940s to 1969, Hartland Plastics, Inc., of Hartland, Wisconsin, produced durable, high-quality horses, dogs, horse-and-rider sets, and other plastic models in a remarkable variety of sizes, shapes, and colors that endeared them to collectors. Hartlands thrilled the children of the 1950s and 1960s who received them as gifts or saved their allowances to buy them. Horse lovers of all ages could appreciate the finely detailed sculpture, imaginative colors, interesting poses, and accurate and graceful gaits. They were realistic, yet pleasantly stylized.

Hartland made two groups of horses. Over 150 different horses of general, riding type accompanied riders or were sold separately between the 1950s and the 1990s. The second group of horses, representing specific breeds, began in 1961. More than 54 shapes of horses depicting 13 breeds and scores of colors, both realistic and fanciful were made, along with 16 dogs.

Hartland Plastics ended model production in 1969, but a second company bought the molds and sold Hartland horses and dogs in the early 1970s. Then, the molds were idle for 10 years until a toy company in Missouri, Steven Manufacturing, and a fourth company made the horses again in the 1980s and 1990s.

Considering the accurate and finely molded detail, such as muscling on the body and ripples in the mane, these models could have been porcelain, rather than mass-produced, plastic "toys." Indeed, their production method permitted a level of quality usually reserved for fine art.

Other Hartland toys destined to become collector items were the standing gunfighters, baseball and other sports statuettes, farm animals, and a farm family. In addition, Hartland Plastics manufactured models for adults, such as plastic religious statuettes and cake-top decorations for weddings, graduations, and other celebrations.

Hartland Plastics did both proprietary and custom molding; that is, it made products sold under its own name, and made components or end products used or sold by other firms, under other names. Between the 1940s and the 1970s, Hartland Plastics had a hand in making figurines for clocks, cosmetics containers, and point-of-purchase displays such as beer and soft drink signs, and other functional, plastic items that do not have the name "Hartland" on them. America was touched by Hartland in more ways than models.

Hartlands were the product of many imaginations and talents, above all, those of the two artists who designed and sculptured all of the original Hartland models, Roger Williams and Alvar Bäckstrand, and the owners, executives, and others who made it all possible.

For Fun Only. Model horses are not investments. Buy them because you like them, and buy only what you like. Be selective. A good collection is one that you enjoy, not necessarily one that is large, "complete," or expensive.

Hartland made two kinds of horse models: (1) horses designed for the rider series (and sometimes sold without riders), and (2) horses of specific breeds ("breed-series" horses).

The "**horses designed for riders**" (HDFR) all have either a hole through the mouth (for a rein to pass through) or have rigid, plastic reins as part of their molded shape. A few examples of the HDFR are shown, for comparison only, in the photo introduction; they are covered exhaustively in a separate book, *Hartland Horsemen*, by Schiffer Publishing, 1999. The breed type of the HDFR is generic.

The subject of this book is the horses of **specific breeds**. They do not have reins or a rein hole (except for a 7.5" high Clydesdale), and they were never sold with riders.

How this Book is Organized

The photo gallery begins with a survey of the Hartland horse molds. Note that some models look alike except for their size. Then, it illustrates one mold (shape) or a group of molds, such as a three-piece family, at a time. Within each mold or family group, the photos are arranged by color, with similar colors adjacent. Very rare models (fewer than 10 made) and Hartland copies, if any, are usually at the end of the mold group. Values are in the photo captions.

(Note that a **mold** is a particular design—shape and size—of horse. A mold is also part of the equipment used to manufacture models. A **model**, collectively speaking, is all the pieces of the same shape, size, and color. A model is also an individual example of its kind. The term "statue" is sometimes used in place of "model." Hartland models were, of course, pre-assembled; they were not model kits (to be built). **Piece** is another term for an individual example of a model in the collective sense.)

The text half of the book is also arranged by molds or family groups, but within each chapter, the models are listed chronologically and grouped by manufacturer, starting with the earliest manufacturer, Hartland Plastics, and proceeding through Durant, Steven, and Paola.

The Four Manufacturers

Hartland's breed series horses began in 1961 and have been made during four decades. They have not been made continuously by a single company, however. Twice since 1969, the product line was sold and traveled to a new manufacturer, and there have been stretches of time when no models were made. This book will abbreviate the four Hartland horse manufacturers as:
(1) **Hartland Plastics**—which made breed series horses from 1961-1969.
(2) **Durant**—the Durant Plastics division of Strombecker Corp. Durant made Hartland-mold horses from 1970-1973. Those horses have the name, "Strombecker/Circle H/Durant," on their package.
(3) **Steven**—Steven Manufacturing Co. produced Hartland horses from 1983-1994. Now called Steven Toys, it is the current owner of the Hartland molds.
(4) **Paola**—Paola Groeber, who bought molded bodies from Steven and painted and marketed them independently from 1987-1990 (with a few also painted in 1986 and in 1991-1992). Paola Groeber did business as Hartland Collectables, Inc., but Steven Mfg. used a similar title (without the "Inc.") on its 1993-1994 catalogs. To avoid confusion, the two manufacturers are designated as "Steven" and "Paola."

Where no manufacturer name is specified, it is correct to assume that it is Hartland Plastics, the original company. The term "Hartland" by itself usually encompasses all four manufacturers; you'll be able to tell from the context.

Distributor vs. Manufacturer. In the 1980s, Paola began as a distributor of Steven models before she became a manufacturer. Her 1985 catalog, 1985 Christmas flyer, and 1986 catalog sold Steven models. Her 1987 catalog had two pages: a page of Steven models (with the #270 Indian set at the bottom) and a separate page of her own products. She told me that the models she sold as test colors at Christmas 1986 and May 1987 were her product, not Steven's. In 1988-1990, she was selling her own models.

How The Model Lists are Organized in the Text

In the text, the information on each model can include: (1) color name; (2) plastic type; (3) production years; (4) manufacturer and catalog number; (5) quantity made; (6) description of the model's appearance.

Color Name. Many of the color names the manufacturers gave the models are accurate and helpful for distinguishing one model from another. However, the four manufacturers used the terms, "bay," "chestnut," and "sorrel" inconsistently, and some models have the same name, but look different. Also, the colors of some models pictured in the 1960s catalogs were never identified. To solve those problems, this book assigns a "nickname" to the model's color, followed by a colon and the actual, catalog name (if there is one). For example, chestnut models from the 1980s-1990s with light-colored (flaxen) mane and tail are nicknamed, "sorrel," to be consistent with the 1960s catalogs. Sometimes, the catalog name is farther back in the model's description.

Plastic Type. The breed series Hartland horses were made in two types of plastic, and all four manufacturers used both types. **Styrene** is a light-weight and relatively brittle plastic. When you gently tap a styrene horse with the side of your fingernail, it sounds "tinny." **Tenite** is familiar as the plastic of Breyer and Peter Stone models. Tenite is actually a brand name used in this book to refer to cellulose acetate plastic, known as "acetate." Tenite (acetate) is heavier and more durable than styrene, and sounds quieter and more substantial when tapped with a fingernail. It is a better quality and more valuable plastic than styrene. In instances of different models having similar colors, being able to discern the type of plastic can help you identify the model.

Quantity Made. The 1960s models were mass produced; the 1970s models and 1980s-1990s Steven models were mass produced, also, but usually in smaller quantities than the 1960s models. All of Paola Groeber's models numbered fewer than 2,000, and most were under 1,000. They could be considered limited editions. Some of the Steven models were limited editions. If no quantity is stated, assume that the model was mass produced.

Description of the Model's Appearance. Horse colors are defined by the body color and color of the mane and tail, with white markings on the face and lower legs being incidental, but white on the body, itself, being significant. For models in this book, the color of the lower legs can be important for telling apart otherwise similar models. The model color descriptions include (1) body color; (2) mane and tail color; (3) the color of the lower legs; and may include, (4) hoof color; and (5) white markings on the face, if any; and (6) the color of the plastic. (If the plastic color is not stated, assume that it is white.) In the lists of models, "mane, tail, lower legs, and hooves" may be **abbreviated** as: "**m-t-ll-h.**" You'll also find, "**m-t-h,**" etc.

Points is the collective term for the mane, tail, and lower legs. When those locations all match each other and contrast with the body, we say that the horse has, for example, "black points" (or brown points or red points). Horsemen never use the term "points" when the m-t-ll are white. However, for convenience sake, I'll describe some models as having "white points" or "beige points," as the case may be. The points do not include the feet (hooves), which may or may not match the lower legs. On some Hartlands, the mane, tail, and hooves all match, but the lower legs do not; the term "points" cannot be used in that case. Instead, the description will read, "black mane, tail, and hooves" (abbreviated as "black m-t-h").

Socks (or stockings) are white markings on the lower legs. Technically, socks are lower; while stockings reach almost to the knees or hocks, but I sometimes use the terms interchangeably. In horse terminology, socks are only white, and black on the lower legs is covered in the term, "black points." However, "black socks," is a more convenient term for models than "black lower legs."

White on the face is a **star** if confined to the forehead, but a **blaze** if it runs the length of the face. A **bald face** has an entirely white front, but the white goes no farther to the sides, in which case it would be an **apron face**. A **snip** is a small white marking in the area between the nostrils. There are additional types of face markings on horses, but this covers most that are found on original finish, mass-produced models.

Original finish models are as the factory finished them; in other words, not retouched or repainted. **Customized** models have had their shape, color, or surface altered in a deliberate effort to change them from their original appearance. Customized models are popular in some collecting circles, and model horse shows have separately-judged classes for them. The models pictured in this book, unless noted otherwise, are original finish.

Colors of Hartland Model Horses

Plastic horses don't always fall neatly into real horse color categories, and it is more to the point that models need color labels that help distinguish one model from another, so buyers and sellers can communicate successfully if the model is sight unseen. However, horse color terms are at the root of model color descriptions. These real-horse colors also describe Hartland horses, with some adjustments that are noted:

There are three common types of "spotted" horse colors: pinto, appaloosa, and dapple.

Pinto—A pinto typically has irregularly-shaped, white areas the size of dinner plates "splashed" on its body.

Appaloosa—An appaloosa typically has spots that are more like the size of 50-cent pieces, and that are solid in color. Appaloosa spots can be dark or light. Some appaloosas have a white area over the hips with small, dark spots on it. That appaloosa pattern is called a "spotted blanket."

Dappled Colors—Everyone has heard of "dappled grey." Dapples are rounded, light areas about the size of 50-cent pieces; they are not solid in color or distinct in shape.

Bay—A horse with a brown or red-brown body and a black mane and tail is usually a bay. The shade of brown or red-brown can vary widely; it is the accompanying, black mane and tail that makes the horse a bay, rather than a chestnut. Hartland also made horses in orange and red, so for purposes of the color nicknames in this book, a **bay model** has a brown, maroon, red, or orange body with a **black** mane and tail.

Chestnut—Chestnuts have many of the same body shades as bays, but lack the black mane and tail. They have a brown or red-brown body and a mane and tail that match the body, are lighter than the body, or are slightly darker than the body, but not black. Some horsemen use the term **sorrel** interchangeably with chestnut, while others consider sorrel a separate color group. Among those who consider sorrel separate, there isn't universal agreement on what it is. In the 1960s, Hartland Plastics used the term "sorrel" for its brown or orange horses with beige or white manes and tails. For purposes of the color nicknames in this book: a **chestnut model** has a brown, maroon, red, or orange body with a mane and tail that **match the body** or are a darker shade than the body, but are not solid black; a **sorrel model** has a brown, maroon, red, or orange body and a mane and tail that are **white, cream, or beige.**

Note: Bay, Chestnut, and Sorrel often appear with modifiers; this book includes models described as: Red Bay, Orange Sorrel, etc.

Palomino—A yellow or golden-yellow horse with a white mane and tail is usually a palomino; likewise for Hartland models. Please note that the mane and tail must be white for a model to qualify as palomino; a model that is yellow-all-over is not a palomino; it should be called "yellow" or "golden-yellow." The all-yellow models are common among the unpainted, 7" and smaller horses.

A yellow horse with a black mane and tail is a **buckskin** (or dun). Hartland buckskins have bodies that are yellow, dusty beige-yellow, golden yellow, or dark golden-yellow (with a black mane and tail).

White Horses. There are four types of white, Hartland horses: (1) unpainted, molded in white plastic; (2) models painted opaque white or Pearl White over white (or sometimes, colored) plastic; (3) models with white bodies and black mane, tail, and stockings (and hooves), which are called **white**

with black points; and (4) models with a white body and gray mane and tail, hooves, and sometimes, gray stockings, which I will call **white with gray mane and tail** or **white-grey**. (Grey spelled with an "e" is a technical term for horses whose color gradually lightens with age, approaching white. Only some horses have that genetic trait; most horses, once they are past their babyhood, remain the same color their whole life, unlike humans, who if they live long enough, all turn "grey" sooner or later.)

Other Horse Colors. Some additional—and complex—horse colors painted by Hartland in the 1980s-1990s, include: **blue roan**—painted as a gray body with black points and black on the head (at least the whole face-front, and sometimes more of the head); **grulla** (blue dun)—like blue roan, but without the black face (a black muzzle does not count); **red roan**—the clearest examples have a pinkish (maroon mixed with white) body with maroon points, and maroon on the head; **"rose grey"**—painted like red roan, but without the maroon head and lower legs; **red dun**—better models have a light orange body and brownish red points; and **"creamy dun"** [light buckskin]—off-white body and black points.

This list of colors doesn't cover all models in the book—after all, some of them are blue or metallic gold, or were deliberately left unpainted in various colors of molded-in plastic—but it goes a long way.

Woodcuts. Nine molds of Hartland horses were designed to look carved from wood. They have a "whittled" surface, and six of them also have engraved "grain lines." In the catalogs, the three without grain lines were called "woodcuts" or "woodgrains," while the others were always called "woodgrains." In this book, I refer to them all as "woodcuts" because it suggests the "carved" texture that all nine of them have. (This contrasts with Breyer woodgrain models, in which the look of wood is merely painted on.) The woodcuts were molded in black, dark red-brown, or tan plastic. Typically, dark stain was hand-rubbed onto the brown and tan models, and grayish highlights were added to the black models, but some woodcuts have no stain or highlights and look fine without it.

Unpainted Models. Hartland horses in the 7" and smaller sizes were often sold in unpainted, solid colors. Unpainted, white models in the 9-11" sizes (and an unpainted, brown Polo Pony) were sold by Paola in the late 1980s. The factory did not usually name the colors of the 7" and smaller, unpainted models. I call the colors: white, ivory-white, black, very dark gray, yellow, golden yellow, cream (very pale yellow), pale taupe (a pale, grayish brown color that some might call bone, putty, malt, off-white, oatmeal, etc.), light gray, red-brown (many would just call this color brown), plum brown (darker than red-brown and with a touch of purple), and cranberry (maroon or burgundy with a hint of magenta; this color is lighter than red-brown and plum brown).

Painted Details. Except for woodcut and unpainted models and two 7" Arabian Stallions from 1968, all Hartland horses have eyes and nostrils painted black, gray, or dark brown. The inner ears are also painted dark on most 1960s horses. The dog models have black eyes and noses. Since these painted details are always present, they are not mentioned in each model description.

Eye whites are found on some 9-11" 1960s models, all Steven models from 1993-1994, and on some examples of Paola's 9-11" models. Of Paola's Christmas 1986 models, the "sample" models had eye whites, but the "test colors" did not. None of Paola's May 1987 models had eye whites except for the black 9" Thoroughbreds. Eye whites are noted in the model lists.

The Molds (Shapes) of Hartland Horses and Dogs

I assigned numbers to the molds and listed them in the order they appear, except that Molds 1-11 are found in *Hartland Horsemen*. Molds 12-65 are the breed series horses, R1-R6 are the resin horses, D1-D16 are the dogs, and E1-E3 are the Extra Horses. An asterisk (*) indicates a mold with two or more minor shape variations. For heights and eras of these models, see the chapters of text for each series.

Horses Designed for Riders:

1. Large Western Champ*
2. Small Western Champ*
3. Chubby*
4. Standing/Walking*
5. Semi-Rearing*
6. Rearing
7. Head-Down Prancer
8. Wrangler Horse
9. Mini Standing/Walking
10. Mini Semi-Rearing
11. Mini Head-Down Prancer

9" Breed ("Individual") Series:

12. Thoroughbred
13. Arabian Mare (Grazing)
14. Five-Gaited Saddlebred
15. Polo Pony
16. Arabian Stallion
17. Weanling Foal
18. Mustang (Rearing)—smooth
19. Mustang (Rearing)—woodcut
20. Tennessee Walker—woodcut*
21. Three-Gaited Saddlebred—woodcut

11" Series:

22. Arabian Stallion
23. Five-Gaited Saddlebred
24. Quarter Horse (Head Tucked)
25. Grazing (Stock Horse) Mare
26. Arabian Mare, Lady Jewel
27. Arabian Foal, Jade

Resin Horses:

R1. American Miniature Horse
R2. Peruvian Paso
R3. Quarter Horse (standing)
R4. Tennessee Walker
R5. Arabian Gelding
R6. Arabian Stallion, Simply Splendid

7" Family Series (and 6" size):

28. Arabian Stallion
29. Arabian Mare
30. Arabian Foal
31. Thoroughbred Stallion
32. Thoroughbred Mare (grazing)
33. Thoroughbred Foal (nursing)
34. Quarter Horse/Appaloosa Stallion
35. Quarter Horse/Appaloosa Mare
36. Quarter Horse/Appaloosa Foal
37. Morgan/Pinto Stallion
38. Morgan/Pinto Mare
39. Morgan/Pinto Foal
40. Tennessee Walker Stallion

41. Tennessee Walker Mare
42. Tennessee Walker Foal
43. Saddlebred Stallion—woodcut
44. Saddlebred Mare—woodcut
45. Saddlebred Foal—woodcut
46. Saddlebred Mare—smooth
47. Saddlebred Foal—smooth
48. Thoroughbred Mare—woodcut
49. Thoroughbred Foal—woodcut
50. Arabian Stallion—woodcut (6" size)

5" Family Series:

51. Arabian Mare
52. Arabian Foal
53. Thoroughbred Mare (grazing)
54. Thoroughbred Foal (nursing)
55. Quarter Horse Mare
56. Quarter Horse Foal

Farm Series & Nativity Set:

57. Farm Horse
58. Farm Donkey
59. Nativity Donkey

Tinymite Horses:

60. Arabian*
61. Belgian*
62. Thoroughbred*
63. Quarter Horse*
64. Tennessee Walker*
65. Morgan (Head Tucked)*

Large Dogs:

D1. Bullet
D2. German Shepherd (Lying Down)

Tinymite Dogs:

D3. Beagle
D4. Black Labrador (Labrador Retriever)
D5. Chesapeake Retriever
D6. Cocker Spaniel
D7. Collie
D8. English Pointer
D9. German Shepherd
D10. German S. H. (Shorthaired) Pointer
D11. Golden Retriever
D12. Irish Setter
D13. St. Bernard
D14. Std. (Standard) Poodle

Farm Dogs:

D15. Adult Farm Dog
D16. Farm Puppy

Extra Horses:

E1. White Horse (on a base)
E2. Bay Budweiser Clydesdale* — only the type with right hind hoof lifted high (not touching the ground at all)
E3. Gold Budweiser 8-Horse Hitch — only the 12" long size (15" long counting the entire sign)

More Than A Dozen Breeds. The horse breeds depicted by Hartland, Molds 12-65, R1-R6, and E1-E3, are:

Arabians (14 molds plus two resins): 11" Stallion, Lady Jewel mold, Jade mold, 9" Stallion, 9" Mare (grazing), 7" three-piece family, 5" two-piece family, 6" woodcut Stallion, and Tinymite; resin Arab Gelding, and resin Arab Stallion.

Thoroughbreds (9 molds): 9" series, 7" three-piece family, 5" two-piece family, 6" two-piece woodcut family, and Tinymite.

Saddlebreds (8 molds): 11" Five-Gaiter, 9" Five-Gaiter, 9" woodcut Three-Gaiter, 7" three-piece family (woodcuts), and 7" two-piece family (smooth).

Stock Horses—Quarter Horses, Appaloosas, and Paints (8 molds plus one resin): 11" series (with head tucked), 11" series Grazing Mare, 7" three-piece family, 5" two-piece family, and Tinymite; resin QH.

Tennessee Walkers (5 molds plus one resin): 9" series woodcut, 7" three-piece family, and Tinymite; TWH resin.

Morgans/Pintos (4 molds): 7" three-piece family, and Tinymite Morgan.

Mustangs—Rearing (2 molds): 9" series smooth, 9" series woodcut.

Draft Horses (one mold, plus two extra, product-sign molds): Tinymite; Single Clydesdale, and 12" long hitch.

Donkeys (2 molds): Nativity Donkey and Farm Donkey.

Others (one of each): Polo Pony (9" series); American Miniature Horse (resin); Peruvian Paso (resin); Saddlebred or Arab type (Farm Horse); Weanling Foal of light-breed type (9" series); Horse of general type: (white horse on a base).

Finding Hartland Models. They're found wherever collectibles are sold, plus in model horse journals. At an auction web site such as eBay, where over 100 Hartland breed-series horses are on sale every day, search for, "Hartland." Unmarked Hartlands might be found under "Breyer," "plastic horse," or "plastic horses." A search for "Hartland Plastics" will turn up few listings because sellers don't use the full name.

Identifying Hartland Horses and Dogs

Mold Marks. Since there are some very close copies of a few of the models, the mold mark can establish the identity. Between the 1940s and 1969, Hartland Plastics used three different mold marks. They are usually located on the right hind leg of the horse, high on the inner thigh. The earliest, Hartland mold mark is an "I" in a diamond with 90 degree angles; in other words, a square that has been rotated 45 degrees. Only one horse in this book has the Diamond "I" mark, which Hartland Plastics used through 1953. (However, some molds that originated prior to 1954 have no mold mark at all, and models produced later than 1953 can have the Diamond "I" mark if their *mold* originated in 1953 or earlier.) Diamonds are a common mold mark on plastic toys, and various diamond marks (usually, with tall or wide, elongated diamonds) belong to companies in Hong Kong and elsewhere. Animals mold-marked, "Hong Kong" or "H.K." were not made by Hartland.

The second Hartland mark is, "Hartland Molded, Hartland, Wisc." It's found on some molds that originated between about 1952 and 1956. It is not found on any breed series horses, but one of my examples of Roy Roger's dog, Bullet, has it. The third mold mark is, "© Hartland Plastics, Inc." Most of the breed series horses (and rider series horses, also) have this mark. Horse molds that originated in 1965-1967 have no brand mark. This was a period when Hartland Plastics was owned by Revlon, Inc., and that may have had something to do with it. Among the dogs, the Tinymite dogs, farm dogs, and Lying Down German Shepherd, which is hollow-bottomed, have no brand mark, either. The absence of a mold mark is not evidence that an item was made by Hartland since many companies have omitted mold marks.

The 1970s horses had the same brand mark as their 1960s counterparts. In the 1980s, a mold mark with the Steven name was added to the 11" series horses, and the mark on the 9" series horses was modified to leave "Hartland" prominent, but diminish the word "Plastics."

Mold Marks
for Breed Series Horses:

Not marked with the Hartland name:
(1) 11" series in the 1960s-1970s (three shapes: molds 22, 23, 24)
(2) 9" Saddlebred woodcut
(3) 7" Tennessee Walker family (1960s and 1980s-1990s)
(4) 7" woodcut Saddlebred Family
(5) 7" smooth Saddlebred set
(6) 7" woodcut Thoroughbred set
(7) 6" woodcut 6" Arab
(8) the farm horse and donkey, nativity donkey, and Tinymites
(9) 11" series Grazing Mare (1980s-1990s)
(10) some examples of 9" woodcut Mustangs and Tennessee Walkers.

Marked, "Hartland/Steven":
(11) 11" series in the 1980s-1990s.

Marked, "Hartland/Steven" on the right leg and "Moody 88" on the left leg:
(12) Lady Jewel and Jade molds (11" series, 1980s-1990s)

Marked, "© Hartland" (with a lump or ridge where the word "Plastics" had been):
(13) 9" series—1980s-1990s.

Marked, "© Hartland Plastics, Inc.":
(14) 9" series—1960s, except for (2), above
(15) 7" series—1960s through 1990s, except for (3), (4), (5), and (6), above
(16) 5" series
(17) some examples of 9" woodcut Mustangs and Tennessee Walkers (but the "Inc." is missing from them).

Assembly Letters. The single letter found on the right or left inner hind leg (or both hind legs) of many Hartland horses was used to identify molded sections to the assemblers; it is not part of the brand mark. The assembly letters on a given horse shape often changed over time because the molds were replaced with duplicates and parts were re-lettered. Most Hartland horses were assembled from six molded parts: left and right sides of the horse and a section for each inner leg. When a mold was replaced, all sections were replaced at the same time.

Describing Hartland Models

Descriptive Title. An Internet auction typically allows space for only a short title for each item. To describe Hartland, breed-series horses when space is limited, list these elements in this order:

1. Brand name: "Hartland." For a Hartland copy, say: "Copy of Hartland..."

2. Size: use the series size—11", 9", 7", 6", 5", or Tinymite (under 3") series—or the actual height; the length is helpful only for identifying low-headed horses, such as grazing mares.

3. Color (or color and catalog number). Don't give only the catalog number since most collectors don't know the models by number. "Unpainted" can be abbreviated as "Unp." Color can also be listed last.

4. Breed. Useful abbreviations include: Thoroughbred = TB, Quarter Horse = QH, and Tennessee Walker = Tenn. Walker or TWH, American Saddlebred = ASB. Other breed names, Morgan, Mustang, etc., should be spelled out.

5. Gender/Age. If the model is identified as a Stallion, Mare, or Foal, mention it.

Western Horses. The above applies to breed series horses. For Hartland western series horses, mention: the brand (Hartland), height and mold shape (there are 11 of them), and whose horse it was (if it belonged to a particular rider) and/or the color.

If unsure of any of the points 2 through 5, above, it is better to omit than to misidentify. However, vague descriptions are annoying: "Brown Hartland Horse" doesn't tell me very much.

Examples of concise, descriptive titles for models:

Hartland 9" Bay Thoroughbred
Hartland 9" ASB — Copper Sorrel
Hartland 9" Tenn. Walker — Walnut Woodcut
Hartland Mustang — Cherry Woodcut
8" Copy of Hartland Mustang
Hartland 7" Gray Arabian Family
Hartland 7" Morgan Mare — Copper Sorrel
Hartland 7" Pinto Stallion
Hartland 5" Bay TB Mare & Foal
Hartland 5" Arabian Mare — Unp. Yellow
Hartland 5" Buckskin QH Mare & Foal
Hartland Dog — Roy Roger's Bullet
Hartland Tinymite Irish Setter
Five Hartland Horses — 5-7" series

The full description of the model should include the condition (noting damage) and the type of plastic (styrene or Tenite) if you are sure. Describe any chipped areas (such as ears), split seams, dirt or marks, missing parts (tail), repairs, factory flaws, etc. If the model needs cleaning, say so. A photograph of at least one side of a Hartland model is essential for an Internet auction. It not only reveals condition, but helps buyers identify the model.

Values

The value of a model depends on two things: its identity and its condition. Not just whether it is a Hartland, but which Hartland model it is, makes a big difference. However, the condition of the particular model is most important of all. Is it chipped, scratched, or scraped? Factors that contribute to different values for different models are: size, type of plastic, scarcity, and of course, the condition of the individual example. All else being equal, larger models and Tenite models are usually more valuable, but all else is seldom equal.

An irony is that, in the 1960s, large quantities were made of popular models of the 9", 7", and the 5" series. Now, those models are common, which cuts down on the demand for them. The number of years that a model appeared in the catalog is also not a reliable measure of how plentiful it is. The Blue-Gray 9" Arabian Mare appears in only the 1964 catalog, yet it was the most common, 1960s color on eBay in 1998-1999. Perhaps it is culled from collections more often because of the unrealistic color, but my guess is that a disproportionately large number of 9" series horses were produced in 1964. I think horse production was high in 1964 and 1965, and not as high in 1966-1968.

Some woodcut horses were sold with felt pads under their feet. I do not add value for that. Nearly all models were manufactured with a small hole on their underside, near the hind legs, so that the model would not collapse inward when it cooled after being molded. The hole is not a flaw or sign of damage, and need not be mentioned by sellers.

An auction price is not a reliable measure of a model's value. Typically, more than one person is willing to pay what a model is worth; to win an auction often requires paying more than it is worth. Bidders who haven't done their homework will often pay too much.

Those accustomed to Breyer horses will need to adjust their thinking. Woodgrain Breyers can be more valuable than regular colors on the same molds because the woodgrain colors were painted for a shorter time. In Hartlands, however, the woodcut models were separate molds, and the Mustang and Tennessee Walker woodcuts, in particular, are quite common. Likewise, Breyer "decorator (non-realistic) colors" often command high prices, but even very common Hartlands are found in "decorator colors," so the term doesn't mean much when applied to Hartlands.

Hartlands are also subject to types of damage not found in Breyer horses. Hartland's molds allowed its horses to exhibit fine, sculptured detail, but also gave them relatively large, hollow spaces in their interior. They can break if stepped on, and they are susceptible to split seams if subjected to 110 degree heat. Less than 1% of 1980s-1990s Hartlands have paint that can get sticky in high heat. In hot weather, a bump or other insult to a 1960s styrene model can show up as melt marks (gouges) in the plastic. In addition to breaks and chips, unpainted Hartlands, not just the painted ones, can sustain scuffs and scrapes that sellers should mention.

Rating Condition and Value. For those who want to be "scientific" about assessing condition and assigning value, here is a method:
(1) note the damage and add up the per cent of damage deductions in the table below.
(2) subtract the total from 100%.
(3) identify the model's condition category (poor to mint).
(4) estimate its value using the value range for that model, which is found in the photo caption.

I use six condition categories, but I only use them to help assign value. Their meanings are not precise, and they are not a substitute for describing the actual qualities, good or bad, of the model's condition or providing a photograph.

Damage to Shape—Deductions:

For each chipped ear: top half gone, 25%; top quarter gone, 20%; tip only gone, 10%.

Bowed leg or splayed foot: 10-15% per leg or foot.

Split seam(s): very visible, 25%; minor, 15%.

Gouge(s), 15%; melt marks on unpainted styrene, 5-20%, depending on size.

Missing parts: the tail, 40%; part of leg, 50%; part of mane chipped off, 15%.

Leg or tail glued back on, 20%-40% depending on the skill of the repair.

Damage to Color/Finish—Deductions:

Rubs, scuffs, or scrapes on painted areas that, in total area (all rubs added together), are: less than the bottom of a pencil eraser, 5%; more than the bottom of a pencil eraser, 10%; more than the area of a dime, 20%; more than a quarter coin, 35%; more than two postage stamps, 50% or more.

Marks (color added): small and few, 10 % for the group; obvious, 25%.

Scuffs/dull spots on unpainted areas, with a total area of: less in area than a postage stamp, 10%; more than a postage stamp, 20%;

Fading of painted areas: extensive, noticeable, 25%.

Yellowed seams or yellow or brown areas from glue spillover (on styrene plastic): 10%-15%.

Obvious yellowing of white plastic where one side of the model is yellowed and the other side is not, 20-35%; even yellowing, 15-25%, depending on severity. Note: pale ivory color is not considered yellowing so long as it is even all over.

Using this system, a model missing an ear tip (-10%) and with several rubs that add up to more than the area of a dime (-20%), and few and small marks (-10%) loses a total of 40% off a perfect condition rating, putting it high in the "average" category.

Condition Category and Value. If the total of the deductions, subtracted from 100% is:

20% or less, the model is in poor condition and worth about a quarter of the "average" value.

21%-40%, the model is in fair condition and worth about half of the "average" value.

41%-60%, the model's condition is average.

61%-80%, very good, and the model is worth midway between the "average" and "excellent" values.

81%-94%, excellent.

95%-100%, near mint to mint; model is worth more than the "excellent" value; perhaps, its value could only be determined at auction.

Models that changed to unnatural colors on their own (e.g., brown paint turned green), might lose 35% of their value. For peeling or melted factory paint on some 1993-1994 models, use the same deductions as for rubs, scuffs, etc.

Customizing of exceptional quality can enhance the value of common models, but non-artistic repaint work and obvious retouch work will decrease the model's value.

Examples of Condition:

Poor: (a) part of a leg is missing and there is other damage, or (b) both ears are half gone and there are rubs, scuffs, or other surface damage in addition, or (c) the shape is intact, but the model is extremely rubbed, scuffed, gouged, greasy/dirty, etc. About 5% of models are in poor condition.

Fair: (a) a broken part was glued back on, or (b) one ear is half or more gone and there's also surface damage (rubs, scuffs, dirt or grease, etc.), or (c) shape is intact but there are many large rubs, scuffs, or stains. About 15% of models are in fair condition.

Average: (a) up to one-quarter of one ear is missing, but there is almost no surface damage, or (b) ears are intact, but there are rubs, scuffs, or marks noticeable on close inspection. Scuffs may be seen in the gloss coat of glossy models and on the surface of unpainted models or unpainted areas of partly-painted models. About 50% of all Hartland models ever made are now in average condition, and average condition is nothing to be ashamed of.

Very Good: No damage to the model's shape, but minor rubs, scuffs, or marks can be seen on the sides of the horse or in a side view of the horse. About 10% of the models are in very good condition.

Excellent: No damage to shape and surface damage is small and more or less limited to the extremities: ears, nose, tail, and hooves, rather than being seen on the sides of the model. About 10% of models are in excellent condition. These models were not played with.

Near Mint: Surface damage is limited to the extremities and is very minor, such as a total of three flaws that are each the size of the head of a pin or smaller. About 8% of models are near mint. They are models that collectors took very good care of.

Truly Mint: These models look new with no handling damage and no fading. Maybe 2% of models are mint.

Mention Flaws. Beyond estimating value, a major purpose of this book's list of model flaws, is to help sellers observe (and report) damage. Collectibles buyers on the Internet appreciate honesty and a lack of outright greed. A detailed description of flaws is interpreted as honesty, and such sellers tend to be rewarded with higher bids than they were expecting for their non-mint objects (if the damage is not extreme).

Box Values. Empty boxes, especially 1980s-1990s boxes, are of some interest to horse collectors, but do not have great value. (We collect horses, not boxes.) A model in a box is not necessarily valuable because, with most styles of Hartland boxes, a model could be removed, played with, and then put back in the box. Models still sealed in a blister package may be mint, but models loose in an unopened box may be rubbed on the ear tips and other locations. Models fastened inside an unopened box may have contact damage from the rubber bands or other fasteners holding them in place. For a box to qualify as being in excellent condition, a minimum requirement is that the cellophane window must be intact and new-looking. Cellophane will be abbreviated as **"cell."**

Molds Are Fine and Models are Plentiful. Rumors of molds being damaged usually come from unsavory sources (sellers trying to inflate values) and should be ignored. Purported mold condition has little bearing on the future of Hartland since molds can always be repaired or replaced. It's been done before, and it can be done again. Since all Hartlands are currently out of production due to management decisions, all of the molds have equal prospects for the future. Their future doesn't affect their past: a tremendous quantity of models has already been produced.

Value Guide Key

Value Ranges. In the illustrated value guide, the range of two numbers under each photo represents the value for that model if it is in **average to excellent condition.** Models in poor or fair condition will fall below the value range, and the value for truly mint models will be higher.

While the flaws that place a model in its condition range can vary, for purposes of this value guide, assume that a model in **average** condition has, at the minimum, **no noticeably-chipped ears.** More than half of one ear being gone would place the model below the value range in the caption.

Where the value range covers a different range of condition than average to excellent, it appears with a code. The abbreviations are: AV = Average, VG = Very Good, EX = Excellent, and NM = Near Mint. For example, VG-NM: $10-$25. Fair and poor condition are not abbreviated.

Collector Models. Paola's models and the 1993-1994 Steven show specials were marketed to adult collectors, so those models are seldom in played-with condition, and the value range given for them is VG-NM. Only a NM value is given for resin models.

How I Determined Values. I determined values from observing the availability/scarcity and auction prices of Hartland models on eBay (on the Internet) every day for over two years and elsewhere over the past 25+ years. I also applied my knowledge of other factors contributing to value, including: size; the quality of the sculpture, plastic type, and manufacturing workmanship; the attractiveness of the colors and finishes; and popularity of specific models for model horse show and general collecting purposes.

A Word About Values. The value ranges in this book are only a guideline. There are no limits to what a seller can ask or a buyer can offer, and prices paid will sometimes be lower or higher than the value ranges in this book.

Most Hartland models are common enough (and sold often enough) that their value does not vary widely. Most of the 7" and smaller horses are worth $3-$10. Most of the 9" horses are worth $10-$30. Many 11" series horses are worth $15-$40. About 95% of Hartland breed-series horses fall into those categories.

The notable exceptions are models which manufacturer Paola Groeber made in small quantities (100 or fewer) and the 11" series horses from the 1960s, which although mass produced, are both difficult to find and prized by collectors. Prices paid for Paola's models—with proof that Paola, not Steven, made them—can exceed the value range in this book. For models made in quantities smaller than 10, values are omitted since there is so little evidence on which to base an estimate of their worth.

The range of selling prices for 11" series horses from the 1960s is wide because price depends so much on condition and on collector enthusiasm. Their values are given in stair-step fashion with a series of numbers for AV, VG, EX, and NM condition. For this handful of rare and attractive old models, the price can exceed the NM figure dramatically in auctions and in certain private sales between collectors or collector-dealers specializing in model horses. However, even models that are valuable when in excellent condition are not worth very much in lesser condition, and 1960s, 11" horses with many large rubs (fair condition or worse) sometimes auction for $15 or less.

Chapter 1: **Hartland Horses at a Glance**

This is a visual introduction to the shapes and sizes of Hartland horses; the colors of each shape are illustrated in later chapters. A separate book, *Hartland Horsemen*, covers the 150+ model horses of the "designed for riders" molds, but their 11 basic shapes are included here.

Hartland horses are found in seven sizes (scales) from 2.5" to 10" high. The sizes are: the 11" series, 9" series, 7" series, 6" Woodcut Arabian (the only horse in its scale), 5" series, Farm set size, and Tinymite. These horses represent specific breeds, and they never had riders.

The Large Champ (9.75" high) and Small Champ (7.5" high), both shown in palomino, were designed for riders. Horses designed for riders have a rein hole through the mouth (or have molded-on bridles and reins); the horses of specific breeds do not have reins or a mouth hole. The Champs are molds 1 & 2.

The Hartland Champs were copied by Breyer Animal Creations' Western Pony and Western Horse, shown here in black pinto and palomino. Breyer was a Hartland competitor; Hartland Plastics did not make Breyer horses. The Champ manes are on the right side of the neck; the Breyer manes are on the left.

Breed series horses include the 7" series Arabian Stallion (top row, 6.75" high); the farm set horse (4.25" high); and 6" woodcut Arabian Stallion. They are molds 28, 57 & 50.

The 9" series Arabian Stallion (mold 16) is 8.75-9" high, and the 11" series Arabian Stallion (mold 22) is 9.5-9.75" high.

The 7" series Arabian Mare (6.25" high) and 5" series Arabian Mare (4.5" high) are alike in stance and conformation. Molds 29 & 51.

The 7" series Morgan/Pinto Stallion (*left*, 6.5" high) and Morgan/Pinto Mare (6.25" high) molds are similar. The stallion's tail attaches to his right hind thigh; the mare's tail attaches to her left hock. Molds 38 & 39.

The 7" series woodgrain Thoroughbred Mare and 7" series Quarter Horse/ Appaloosa Stallion are both 6.25" high. Molds 48 & 34.

The heights of the 7" series Quarter Horse/Appaloosa Mare (5.5") and 5" series Quarter Horse Mare (4.25") were measured at the highest part of their arched necks. Molds 35 & 55.

The 11" series Quarter Horse (and Appaloosa) mold, is 7.75-8" high at the highest point of its arched neck. Mold 24. *Courtesy of April and Jon Powell.*

The 5" series and 9" series Head-Down Prancers, 4" and 6.25-6.5" high, respectively, were designed for riders, but were sometimes sold without riders. Molds 11 & 7.

Hartland made two Five-Gaited American Saddlebreds. The 11" series Five-Gaiter is 9.5-10" high and the 9" series Five-Gaiter is 8-8.25" high. Molds 23 & 14.

The 9" series and 7" series Three-Gaited Saddlebred Stallions are similar in build. The 9" Three-Gaiter is 8.5" high; the 7" series Three-Gaited Saddlebred Stallion is 6.5" high. Molds 21 & 43.

The 7" series mares include: Tennessee Walker Mare (top row, 5.75" high); 7" series smooth Saddlebred Mare (6.5" high), and the woodcut 7" series Saddlebred Mare (6.25" high). Molds 41, 46 & 44.

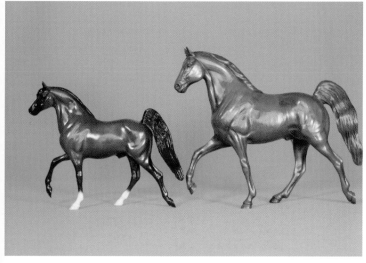

The 7" series Tennessee Walker Stallion (6.5" high) and 9" series Tennessee Walker (8" high) are both gliding along at a running walk. Molds 40 & 20.

The 9" series Thoroughbred (7.5-7.75" high) and 7" series Thoroughbred Stallion (over 6.25" high) are both walking. Molds 12 & 31.

The rider series includes "standing/walking" horses in three sizes: the 9" series S/W horse (8-8.25" high), 5" series S/W horse (4.5" high), and 7" series Wrangler horse (6.25" high). The 9" size includes five mane and tail variations. All three sizes were sometimes sold without riders. Molds 4, 9 & 8.

The "Chubby" horse (8-8.25" high) and Rearing horse (9.5" high to the tip of its nose) were designed for, and almost always sold with, riders. Two mane-tail-bridle variations are found in the Chubby; copies of the Chubby horse are found, too. Molds 3 & 6.

The 9" series Mustangs are found in smooth (*left*, 9-9.25" high) and woodcut (textured) versions. The woodcut Mustang is 9.25" high. Molds 18 & 19.

Two, Semi-Rearing Horses were designed for, and usually sold with, riders. The 9" series Semi-Rearing horse (8.25-8.5" high) includes three mane and tail variations; the 5" series Semi-Rearing horse (4.25" high) had no shape variations. Molds 5 & 10.

The Lady Jewel and Jade molds are part of the 11" series. Lady Jewel is 8" high at the highest point of her mane; Jade is 6.25." Both are Arabians. Molds 26 & 27.

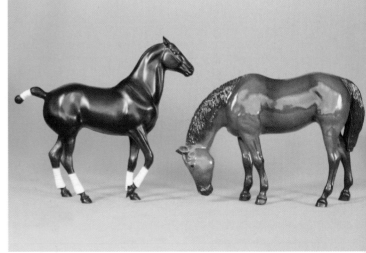

The 9" series Polo Pony (8-8.25" high) wears molded-on bandages. The 11" series Grazing Mare mold is 6.5" high at the highest point—the withers. Molds 15 & 25.

Grazing Mares are also found in the: 9", 7", and 5" series. The 9" series Arabian Mare is 5.75" high, the 7" series Thoroughbred Mare is 4.75" high; and the 5" series Thoroughbred Mare is 3.75" high. Heights were taken at the highest point—the withers. Molds 13, 32 & 53.

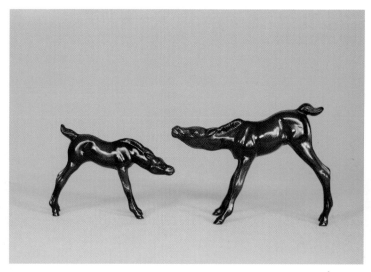

Thoroughbred Foals in the 5" series (2.75" high), and 7" series (3.5" high) are nursing. They were measured at the highest part of their tails. Molds 54 & 33.

The 9" series Weanling Foal (6" high), 7" series Arabian Foal (4.75" high), and 5" series Arabian Foal (3.5" high) are standing still. Molds 17, 30 & 52.

The 7" series, woodcut Thoroughbred Foal (*left*, 4.5" high) and 7" series Tennessee Walker Foal (4.75" high), share the same pose, but differ in size, texture, and color. Molds 49 & 42.

There are two, 7" series Saddlebred foals: the woodcut Saddlebred Foal (4.5" high), and the smooth Saddlebred Foal (5" high). Molds 45 & 47.

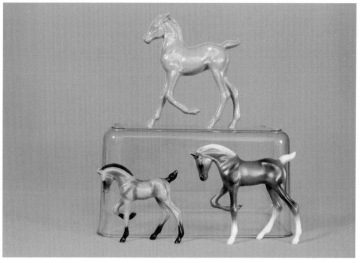

Foals with a forefoot lifted high are the 7" series Quarter Horse/Appaloosa Foal (*top row*, 4.5" high); 5" series Quarter Horse Foal (*lower left*, 3" high); and 7" series Morgan/Pinto Foal (3.75" high). Molds 36, 56 & 39.

The Tinymite Horses, 2.25-2.75" high, are: (top row)—Tennessee Walker, Arabian, and Belgian; (bottom row) — Morgan, Quarter Horse, and Thoroughbred. Each mold has a slight variation; the ones shown are the more common, "joined-ear" type. Copies are also found. Molds 60-65.

The Farm set donkey is 3.75" high; the Nativity set donkey is 4.5" high. Molds 58 & 59.

This 4.75" white horse on a 1.25" base (almost 6" high overall) was an early (pre-1954) item manufactured by Hartland Plastics. Mold E1. *Model, courtesy of Sande Schneider.*

Hartland Plastics manufactured some of the 7.5" trotting, Budweiser Clydesdales (mold E2), but many Clydesdales for beer signs were not made by Hartland. Hong Kong versions of the Clydesdale are common.

In the 1980s and early 1990s, limited-edition resins, such as this Walking Arabian Gelding (8.75" high), were sold under the Hartland name. The sculptor was Linda Lima. Mold R5. *Model, courtesy of Bobbie Mosimann; photo by Michelle Grant.*

"Simply Splendid," the Arabian Stallion resin (9.25" high), was sculpted by Kathleen Moody. Mold R6. *Courtesy of Shirley Ketchuck.*

The Peruvian Paso resin model (about 9.5" high) was also sculpted by
Kathleen Moody. Mold R2. *Courtesy of Tina English-Wendt.*

The Tennessee Walker resin (5" high) was sculpted by Carol Gasper. Mold
R4. *Courtesy of Shirley Ketchuck.*

The Miniature Horse resin, 4.75" high, was a Kathleen Moody sculpture.
Mold R1. *Courtesy of Shirley Ketchuck.*

The Quarter Horse resin (5" high) was a Carol Gasper sculpture. Mold R3.
Courtesy of Carol Gasper.

The 11" and 9" series models from the 1960s-1970s are slightly smaller than their 1980s-1990s counterparts. In the 7" series though, the older models are the same height or slightly taller. In each pair illustrated, the 1960s models are at *left*; 1980s-1990s models are at *right*.

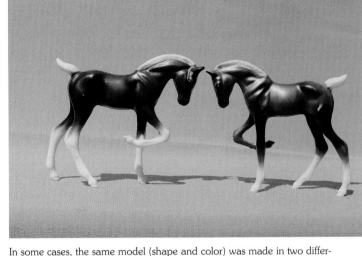

In some cases, the same model (shape and color) was made in two different types of plastic, such as these 1960s, 7" series, copper sorrel Morgan Foals in styrene (*left*) and Tenite (*right*).

To tell styrene from Tenite (acetate) models, tap them gently with your fingernail. The styrene models, which are more light-weight and brittle, sound "tinny"; the Tenite models sound quieter, but more substantial. This copper sorrel Morgan stallion is the less common, Tenite version.

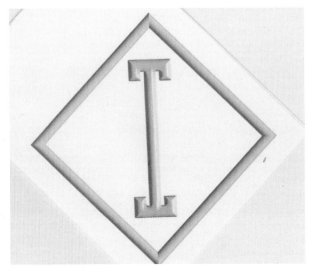

The "Diamond I" mold mark is found on some of the earliest Hartland models. The "I" stood for Iola Walter, co-founder of Hartland Plastics. It's enlarged here from its actual size of about one-quarter inch. *Computer-aided art, courtesy of Michael Mueller.*

The 9" Series Thoroughbreds (mold 12) measure 7.5" to 7.75" high.

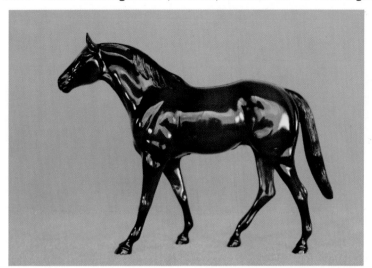

Maroon Bay 9" Thoroughbred (with shadings and gloss coat), Tenite, Hartland Plastics #873, 1963-1967. $10-20.

Brown 9" Thoroughbred, styrene, Steven #231, "Mahogany Bay," 1984-1986. *Model, courtesy of Terry Davis.* $12-22.

Clay Bay 9" Thoroughbred (with body color molded in), Tenite, Hartland Plastics #8001, 1968 only. *Courtesy of Shirley Ketchuck.* $12-22.

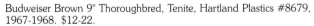

Budweiser Brown 9" Thoroughbred, Tenite, Hartland Plastics #8679, 1967-1968. $12-22.

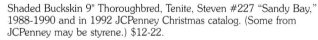

Shaded Buckskin 9" Thoroughbred, Tenite, Steven #227 "Sandy Bay," 1988-1990 and in 1992 JCPenney Christmas catalog. (Some from JCPenney may be styrene.) $12-22.

Clear Buckskin 9" Thoroughbred, Tenite, Paola # 461, "Sandy Bay," 1988-1990. VG-NM: $25-35.

Metallic Gold Bay 9" Thoroughbred, Tenite, Paola #462, 1989-1990. *Courtesy of Traci Durrell-Khalife*. VG-NM: $35-50.

Light Sorrel 9" Thoroughbred, Tenite, #460L, an accidental variation of Paola #460 "Charcoal" (Dark Sorrel), 1988, about 30 made. VG-NM: $40-60.

Dark Sorrel 9" Thoroughbred, Tenite, Steven #228 and Paola #460 (both "Charcoal"), both 1988-1990. Steven: $15-25; Paola (VG-NM): $25-35.

Chestnut 9" Thoroughbred, Tenite, Steven #230, "Rapid Delivery," 1993-1994. *Courtesy of Sandy Tomezik*. $15-30.

"Deep, Rich Chestnut" 9" Thoroughbred, styrene, Paola #331C, May 1987 test color, 100 made. *Courtesy of Peggy Howard.* VG-NM: $30-45.

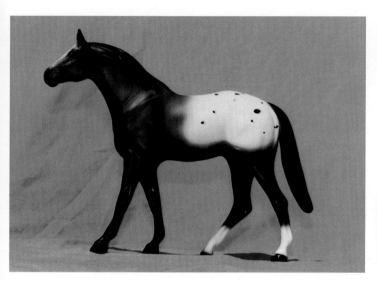

Bay Appaloosa 9" Thoroughbred, Tenite, Steven special for West Coast Model Horse Collector's Jamboree, 1992, 200 made. *Courtesy of Judith Miller.* VG-NM: $35-50.

Chestnut Appaloosa (Tenite, using the 9" Thoroughbred mold); Paola made five in the late 1980s. *Courtesy of Tina English-Wendt.*

Black 9" Thoroughbred, styrene, Paola #331D, May 1987 test color, 18 made. *Courtesy of Judy Renee Pope.* VG-NM: $30-45.

Black Dapple Appaloosa 9" Thoroughbred, Tenite, Steven #202-20 special for Cascade Models, 1994, 50 made. *Courtesy of Daphne R. Macpherson.* VG-NM: $35-65.

"Scammer," Dapple Grey 9" Thoroughbred, Tenite, Paola special for Black Horse Ranch, 1990, 256 made; sold with black Weanling Foal, "Black Embers." VG-NM: $30-45.

Dark Dapple Grey 9" Thoroughbred, styrene, Paola #331A, May 1987 test color, 50 made. Another dark dapple grey 9" Thoroughbred—with a bald face—was a 1990 show special of about 24 in Tenite. *Courtesy of Laura Pervier*. VG-NM: $30-45 (styrene);$35-50 (Tenite).

Light Dapple Grey 9" Thoroughbred, styrene, Paola #331B, May 1987 test color, 50 made. *Courtesy of Sandy Tomezik*. VG-NM: $30-45.

White 9" Thoroughbred (with gray mane and tail), Tenite, Hartland Plastics #8001, 1967 only. $15-35.

The 9" Series Grazing, Arabian Mares (mold 13) measure 5.75" high at the highest point, the withers.

Blue 9" Thoroughbred (with white points), Tenite, Paola, one was given as a prize at a model show in St. Louis in July, 1990. *Courtesy of Shirley Ketchuck*.

Rusty Grey 9" Arabian Mare, Tenite, Steven #221 "Chestnut Flea Bit" (name: "Fair Maiden"), 1993-1994. $15-30.

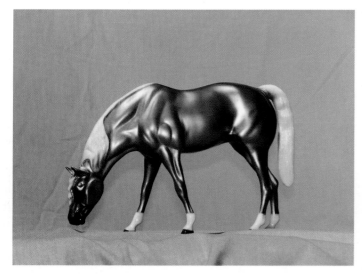

Metallic Copper 9" Arabian Mare, Tenite, Paola #453, 1990 only. *Courtesy of Judith Miller*. VG-NM: $30-60.

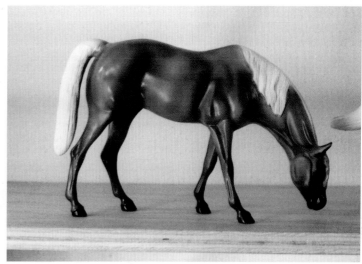

Light Sorrel 9" Arabian Mare, styrene, Paola #339D, "Light Chestnut," May 1987 test color, 45 made. VG-NM: $25-45.

Buckskin 9" Arabian Mare, styrene, Steven #241, 1985-1986. *Courtesy of Daphne R. Macpherson*. $12-22.

Chestnut 9" Arabian Mare, Tenite, Paola #451, 1988-1990, and Steven #240, 1988-1990 and for 1992 JCPenney 1992 Christmas catalog. Some from JCPenney could be styrene. *Courtesy of Eleanor Harvey*. Steven/ JCP: $15-25; Paola (VG-NM): $25-35.

Light Bay 9" Arabian Mare, Tenite, Paola #452, "Light Bay," 1989-1990. VG-NM: $25-40.

1980s Shaded Bay 9" Arabian Mare, styrene, Paola #339C, "Light Bay," May 1987 test color, 45 made. *Courtesy of Maggi Jacques*. VG-NM: $30-45.

1960s Shaded Bay 9" Arabian Mare (painted over white plastic, and not highly glossy), Tenite, Hartland Plastics #8679, 1968 only. $12-22.

Unshaded, Maroon Bay 9" Arabian Mare with dark, brown-red body color molded in (and no shadings), Tenite, Hartland Plastics #8001, 1968 only. $12-22.

This Steven #249 Dapple Grey 9" Arabian Mare (Tenite, 1988-1990), has very large dapples. $15-25.

Typical, Steven #249 Dapple Grey Arabian Mares (Tenite, 1988-1990) have small dapples, a very dark face and dark-gray-to-black knees and hocks. This one has pale hind feet. $15-25.

This Steven #249 Dapple Grey 9" Arabian Mare (Tenite, 1988-1990) clearly exhibits the blocky, white area on the tail where the model was held by a clamp during painting. $15-25.

Dapple Grey, 9" Arabian Mares by Paola (#450, Tenite, 1988-1990) have more subtle shadings on the face, and smoothly blended color on the tail. They varied: the model at left is darker and has one pink hoof; the mare at right has four black hooves and delicate knee-hock-mane shadings. *Courtesy of Eleanor Harvey.* VG-NM: $25-40 each.

With a silvery finish and in styrene plastic, the Light Dapple Grey (Pearled) 9" Arabian Mare is Paola #339C, a May 1987 test color; 31 were made. *Courtesy of Eleanor and Shay Goosens.* VG-NM: $35-50.

The Light Dapple Grey (Pearled) 9" Arabian Mares in styrene with silvery finish (Paola #339C, May 1987 test run of 31) varied a bit; this one has a very dark mane and tail: *Courtesy of Eleanor and Shay Goosens.* VG-NM: $35-50.

Dark Dapple Grey (Pearled) 9" Arabian Mare, styrene, Paola #339A, May 1987 test color, 19 made. *Courtesy of Eleanor Harvey.* VG-NM: $35-50.

Blue-Gray 9" Arabian Mare, Tenite, Hartland Plastics #876, only in 1964 catalog, but common. $12-22.

Dove Gray 9" Arabian Mare (with black points and no dapples), styrene, Steven #239 "Light Grey," 1984-1986. *Courtesy of Eleanor and Shay Goosens.* $10-20.

White 9" Arabian Mare with black points, Tenite, Hartland Plastics #876, 1963-1964. *Courtesy of Judith Miller.* $14-26.

White 9" Arabian Mare with gray mane and tail, Tenite, Hartland Plastics #8001, 1967. $15-30.

Pearl White 9" Arabian Mare (painted over white plastic), Tenite, Hartland Plastics #8679, 1967. *Courtesy of Judith Miller.* $35-75+.

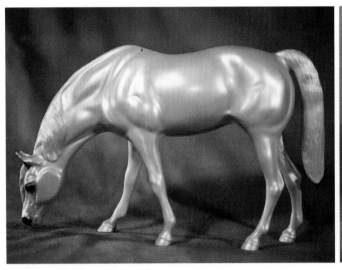

The #8679, Pearl White 9" Arabian Mare often photographs as a deep, creamy white. This example has white plastic underneath its pearl white paint. *Courtesy of Sandy Tomezik.* $35-75+.

Some #8679 Pearl White 9" Arabian Mares were painted pearl white over golden-yellow plastic (*right*). Compare with the Pearl White model painted over white plastic (*left*). *Courtesy of Karen Oelkers.* $35-75+.

The 9" Five-Gaited Saddlebred (mold 14) is 8" to 8.25" high.

Pearl White 9" Arabian Mare, Tenite, with semi-gloss finish, one of only two this color painted by Paola in 1990. *Courtesy of Eleanor Harvey.*

Copper Sorrel 9" Five-Gaiter, Tenite, Hartland Plastics #881, 1963-1966. This example has the "frosted" look, which is a duller orange and more matte, with the metallic/pearled appearance more obvious. $10-22.

Copper Sorrel 9" Five-Gaiters (#881, 1963-1966) in their two variations: "frosted" (right), and "deep" (deeper orange and more glossy, with a less metallic/pearled appearance). $10-22 each.

Orange Sorrel 9" Five-Gaiter, styrene, Steven #235 "Sorrel," 1984-1986. *Courtesy of April and Jon Powell.* $10-20.

Red Sorrel—Matte Finish, 9" Five-Gaiter, styrene, a 1984 mistaken version of Orange Sorrel, Steven #235; about 100 were made in the matte finish. *Courtesy of Carla S. Clifford.* $14-26.

Red Sorrel—Glossy, 9" Five-Gaiter, styrene, a 1984 mistaken version of Steven #235 Orange Sorrel; about 100 were made in the glossy finish; their distributor was Paola Groeber. *Courtesy of Daphne Macpherson.* $14-26.

1960s Red Bay 9" Five-Gaiter, Tenite, Hartland Plastics #8679, 1967-1968. Note apple red color, black shadings, and the four white stockings. *Courtesy of Susan Bensema Young.* $15-30.

The 1980s Bay 9" Five-Gaiter (Tenite; "Light Bay," Paola #433 and Steven #244, both 1988-1990) has black on the legs and muted, reddish color. The amount of shadings varies; this example has few or no shadings on the body. *Courtesy of Laura Pervier.* Steven: $15-25; Paola (VG-NM): $25-35.

This 1980s Bay 9" Five-Gaiter is a deeper red and has more dark shadings than its fellow Tenite, "Light Bay," 1988-1990, Paola #433 and Steven #244 models. This example was painted by Paola. *Courtesy of Laura Pervier.* VG-NM: $25-35.

Liver Chestnut 9" Five-Gaiter, styrene, Paola #335, a May 1987 test color, 72 made; has a blaze. *Courtesy of Eleanor and Shay Goosens.* VG-NM: $30-45.

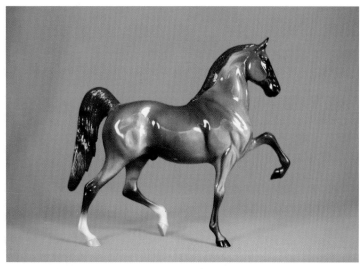

Cinnamon (Red Roan) 9" Five-Gaiter in styrene (Glossy), Steven special for 1991 JCPenney Christmas catalog; common (20,000 made); stallion anatomy was added to this mare mold. $12-25.

Red Roan 9" Five-Gaiter in Tenite with low contrast between body and points, Steven #243, 1988-1990. $15-35.

Red Roan 9" Five-Gaiter in Tenite with more contrast between body and points, Steven #243, 1988-1990. (Model was purchased at a Wal-Mart store in November 1988.) Note that the entire head is dark. *Courtesy of Susan Bensema Young.* $15-35.

Red Roan 9" Five-Gaiter (Tenite) with high contrast between body and points, Paola #432, 1988-1990. The model's body actually looks more grayish-pink than this, and the points look maroon. The face, but not the whole head, is maroon. VG-NM: $25-50.

Red Roan 9" Five-Gaiter with high white stockings (Tenite); Paola made two like this, c.1987. Also note white on face, and high contrast between body and points. *Courtesy of Tina English-Wendt.*

Metallic Gold 9" Five-Gaiter, a 1990-1991 Paola, a show special color; most had pink hooves, but 12 had black hooves; perhaps, so many as 50 were made. *Courtesy of Laura Pervier.* VG-NM: $35-65.

Palomino (Semi-Glossy) 9" Five-Gaiter, Tenite, Paola #431, "Dapple Palomino," 1987-1990. VG-NM: $30-45.

Palomino (Matte Finish) 9" Five-Gaiter, Tenite, Paola #431, "Dapple Palomino," 1987-1990. *Courtesy of Laura Pervier.* VG-NM: $30-45.

Oatmeal Dun 9" Five-Gaiter (with malt/oatmeal/pale taupe body color molded-in), Tenite, Hartland Plastics #8001, 1968 only; mane, tail, and hooves are dark brown. *Photo, courtesy of James W. Truitt; model, courtesy of Sandra J. Truitt.* $15-40.

In bright sunlight, the Oatmeal Dun 9" Five-Gaiter can look the color of cream or buttermilk. *Courtesy of Judy Miller.* $15-40.

Pearled White 9" Five-Gaiter, Tenite, Paola #431P, a 1988 special run color. VG-NM: $30-50.

White 9" Five-Gaiter with gray mane and tail, Tenite, Hartland Plastics #8001, 1967. $15-35.

"Twilight Moon," Blue Roan 9" Five-Gaiter, Tenite, Steven #222, 1993-1994. $15-35.

Blue Roan (Pearled) 9" Five-Gaiter, Tenite, Paola #430P, a 1988 special run; they varied a bit, and this is a lighter blue roan. *Courtesy of Laura Pervier.* VG-NM: $30-50.

The darker version of the Blue Roan (Pearled) 9" Five-Gaiter, Tenite, Paola #430P 1988 special run color. *Courtesy of Laura Pervier.* VG-NM: $30-50.

Light Dapple Grey 9" Five-Gaiter, Tenite, Paola #430, as seen in the 1988 and 1989 catalogs. VG-NM: $30-50.

Dark Dapple Grey 9" Five-Gaiter; Tenite, Paola #430, with darker body color seen in the 1987 and 1990 catalogs. *Courtesy of Eleanor Harvey.* VG-NM: $30-50.

Black 9" Five-Gaiter with silver mane and tail and blue undertones; Tenite, a Paola special painted in 1990, but not sold until December 1994; 25 were made. VG-NM: $35-65.

Black 9" Five-Gaiter, Tenite, Paola #434, "Raven Black," 1990 only. VG-NM: $35-60.

Black Pinto 9" Five-Gaiter, Tenite, Steven #222-22, "CJ Kaleidoscope," a 1994 special of 225 for Cheryl Monroe's model show, Michigan. VG-NM: $35-65.

Left side of Black Pinto 9" Five-Gaiter, "CJ Kaleidoscope."

Black Dappled 9" Five-Gaiter (with white dapples and white mane and tail); Tenite, Steven #222-21, "Zest for Living," a 1994 special of 200 for Debbie Buckler's model show, California. VG-NM: $20-40.

Black Dappled 9" Five-Gaiter with black points; Tenite; Debbie Buckler customized 15 of the "Zest for Living" models (by painting their white points black) and called them, "Starry Night." VG-NM: $20-40.

Red Dun (Taffy Tan) 9" Five-Gaiter, a "mistake" rescued from Paola's trash bin, c.1985; there were three this color. *Courtesy of Tina English-Wendt.*

Palomino test color 9" Five-Gaiter by Paola, styrene; May 1987, three made; has brownish red shadings over yellow base color. *Courtesy of Eleanor and Shay Goosens.*

Copper 9" Five-Gaiter with white points; Tenite, by Paola, two made, 1990. *Courtesy of Daphne R. Macpherson.*

Silver 9" Five-Gaiter with white points; Tenite, by Paola; one or two were made, 1990. *Courtesy of Paola Groeber.*

Wedgewood (Blue) 9" Five-Gaiter; Tenite, by Paola; one of 12 in this color, 1989. *Courtesy of Eleanor Harvey.* VG-NM: $55-125+.

The purple toy at left is a copy of the Hartland 9" Five-Gaiter (shown in 1960s white-grey). A film container cap helps the Hartland model stand. *Purple toy, courtesy of Cecile Bellmer.* Toy: $3-8.

The 9" Series Polo Pony (mold 15) is 8" to 8.25" high. Since there is only one scale of Polo Pony, the size designation, 9", is not essential and is omitted from each caption.

1960s Dark Bay Polo Pony, Tenite, Hartland Plastics #883, 1964-1966. $18-35.

Rich, Dark Bay Polo Pony, styrene, Paola #336B, May 1987 test color, "Blood Bay," 92 made. Note dark shading on muzzle. *Courtesy of Eleanor Harvey.* VG-NM: $35-50.

1980s Seal Brown Polo Pony, styrene, Steven #236, 1984-1986; the points are black, but the muzzle is brown; there are no body shadings. *Courtesy of Maggi Jacques.* $12-22.

Bright Chestnut Polo Pony, Tenite, Paola #218S, 1989 special of 100; some were re-sold by Your Horse Source, 1990, as #H218, "Chestnut." VG-NM: $30-60.

Palomino Polo Pony, Tenite, Paola #217S, "Golden Palomino," 1989 special run of 100; some were re-sold by Your Horse Source, 1990, as #H217, "Golden." *Courtesy of Susan Bensema Young.* VG-NM: $30-60.

Buckskin Polo Pony with dark blue bandages; Tenite, Paola painted three c.1990 and sold them in 1994. *Courtesy of Laura Pervier*.

Buckskin Polo Pony with red bandages; Tenite, Hartland Plastics #8679, only in late 1967 or early 1968 catalog. *Courtesy of Laura Pervier*. $25-45.

Dark Dun Polo Pony with dark blue bandages and shoulder and leg stripes; Tenite, Paola painted three c.1990 and sold them in 1994. *Courtesy of Shirley Ketchuck*.

"Creamy Dun" Polo Pony with royal blue bandages; Tenite, Steven #219, "Quick Twister," 1993-1994. $15-30.

Collector Debbie Buckler sold a "color enhanced" version of the Steven #219 Creamy Dun Polo Ponies; she added dark shadings and lightened the blue bandages. *Courtesy of Laura Pervier*. VG-NM: $20-35.

Red Roan Appaloosa Polo Pony, Tenite, Paola #471 and Steven #228, "Leopard Appaloosa," 1988-1990. Paola's have hardly any reddish color on the knees/hocks and faint shading on the face. *Courtesy of Eleanor & Shay Goosens*. Steven: $15-30; Paola (VG-NM): $25-40.

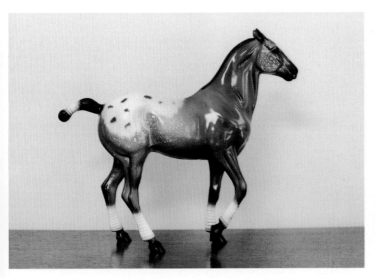

Bay Blanket Appaloosa Polo Pony, styrene, Paola #336A, May 1987 test color, 60 made. *Courtesy of Eleanor Harvey.* VG-NM: $25-45.

Black Appaloosa Polo Pony with red bandages, Tenite, Steven #219-20, a 1993 special of 200 for Debbie Buckler's model show, California; of the 200, 16 were actually brown-black; also, Debbie repainted the wraps purple on ten of the 200. *Courtesy of Sandy Tomezik.* VG-NM: $30-60 each.

Dapple Grey Polo Pony with black bandages, styrene, Paola 3236A, a Christmas 1986 test color, 13 made. *Courtesy of Laura Pervier.* VG-NM: $35-50.

Black Polo Pony, styrene, Steven #236S, a special of 1,000 available in 1984-1986. *Courtesy of Eleanor Harvey.* VG-NM: $18-35.

Metallic Blue Polo Pony, Hartland Plastics #8679, 1967 only. *Courtesy of Eleanor Harvey.* $25-65.

Blue Roan Polo Pony, Tenite, Paola #470 and Steven #229, both 1988-1990, and in Steven's Three-Mare set in the 1992 JCPenney Christmas catalog, of which some could be styrene. Steven/JCPenney: $15-30; Paola (VG-NM): $25-35.

The Blue Roan Polo Ponies are found in lighter and darker versions. The darker version was by Paola, but the lighter version was made by both Paola and Steven. *Courtesy of Peggy Howard.* Darker version (VG-NM): $25-35.

White Polo Pony with gray mane and tail, Tenite, Hartland Plastics #8001, 1967 only. *Model, courtesy of Jackie Himes; photo, courtesy of April and Jon Powell.* $15-40.

Blue Polo Pony with white points, Tenite, by Paola, who gave one as a prize at a St. Louis model show, 1990. *Courtesy of Paola Groeber.*

Pearl White Polo Pony with Wedgwood blue bandages and dark shading around the nose and eyes, Tenite, by Paola, one of only two made, 1989. *Courtesy of Eleanor Harvey.*

The 9" Series Arabian Stallion (mold 16) is 8.75-9" high.

The Polo Ponies in Dark Bay, Bright Chestnut, and Creamy Dun have an "arrow blaze" face marking, but the Blue Roan (*second from left*) has a solid-black face. (The Creamy Dun's blaze did not photograph well.)

Copper Sorrel 9" Arabian Stallion, Tenite, Hartland Plastics #8679, 1967-1968. $18-35.

The left side of the Copper Sorrel 9" Arabian stallion

Chestnut 9" Arabian Stallion, Tenite, Paola #421 and Steven #245, both 1988-1990. Steven: $15-30; Paola (VG-NM): $25-35.

Light Bay 9" Arabian Stallion, Tenite, Paola #420, 1988-1990. $25-40.

Glossy Dark Bay 9" Arabian Stallion, styrene, a Christmas 1985 Steven special for Paola Groeber; 32 made; white markings vary; they are hand-painted. *Courtesy of Sandy Tomezik.* VG-NM: $30-45.

1960s Sorrel 9" Arabian Stallion, Tenite, Hartland Plastics #884, 1964-1966. $15-30.

1980s ("Bright") Sorrel (Glossy) 9" Arabian Stallion, styrene, Steven #237, 1984-1986. $13-23.

Glossy Dark Chestnut 9" Arabian Stallion, styrene, a total of 640 produced by Steven: 140 as a Christmas 1995 special for Paola (#237C), and 500 as an October 1986 special for Black Horse Ranch. Models for Paola had masked-off white socks; socks for the BHR models were hand painted. *Courtesy of Peggy Howard.* VG-NM—BHR: $20-35; Paola: $25-40.

Rose Grey (Pearled) 9" Arabian Stallion, Tenite, Paola 3421P, a 1988 special run. VG-NM: $30-50.

Clay Bay 9" Arabian Stallion, Tenite, Hartland Plastics #8000/8001, 1968. $15-40.

Metallic Gold Bay 9" Arabian Stallion, Tenite, a 1990 Paola model, very few made. *Courtesy of Peggy Howard.*

Palomino 9" Arabian Stallion, Tenite, Paola, a show special color dated September 1989, February 1990, or March 1990; probably 50 or more made. *Courtesy of Eleanor Harvey.* VG-NM: $30-55.

This Dapple Palomino (*left*) 9" Arabian Stallion (Tenite, by Paola, dated 6/1990) was donated to a show in Virginia; dapple palomino was also a February 1990 special of 20-30 for a West Coast model show. The solid palomino (*right*) is more common. *Courtesy of Peggy Howard.* Dappled (VG-NM): $35-60

Pearl White 9" Arabian Stallion, Tenite, Paola #420P, a 1988 special. *Courtesy of Eleanor Harvey.* VG-NM: $30-50.

White 9" Arabian Stallion with gray mane and tail, Tenite, Hartland Plastics #8001, 1967. *Courtesy of Eleanor Harvey.* $20-40.

Some of the White 9" Arabian Stallions (Tenite, Hartland Plastics #8001, 1967) have a black mane and tail, instead of gray. *Courtesy of Eleanor Harvey.* $20-40.

JCPenney Dapple Grey 9" Arabian Stallion in styrene (Glossy), part of the Steven #278 Horse Assortment (Three-Stallion set), a special for the 1991 JCPenney Christmas catalog; common (up to 20,000). *Courtesy of Eleanor Harvey.* $10-22.

Dapple Grey 9" Arabian Stallion in Tenite, Paola #422 and Steven #246. This example, with minimal dark shadings on the face, knees, and hocks and color well blended on the tail, is by Paola. *Courtesy of Eleanor Harvey.* Steven: $15-25; Paola (VG-NM): $25-35.

Light Dapple Grey (Pearled) 9" Arabian Stallion, styrene, Paola test color #237S, Christmas 1986, 35 made. *Courtesy of Traci Durrell-Khalife.* VG-NM: $35-60.

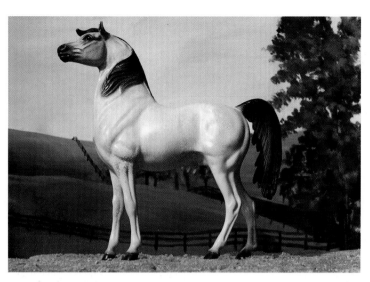

Dapple Grey (Pearled) 9" Arabian stallion with dark gray mane and tail, Paola test color #337B, May 1987, 25 produced. *Courtesy of Eleanor & Shay Goosens.* VG-NM: $35-60.

Dark Dapple Grey (Pearled) 9" Arabian Stallion, styrene, Paola test color #237S, Christmas 1986, 58 made. *Courtesy of Judy Renee Pope.* VG-NM: $35-55.

Raven Black 9" Arabian Stallion, Tenite, Steven #224, "Desert Prince," 1993-1994. $15-30.

Glossy Black 9" Arabian Stallion, styrene, 640 produced by Steven: 140 for Paola Groeber, Christmas 1985, #237B; and 500 for Black Horse Ranch, October 1986. All are solid black except the BHR models have *eye* whites. *Courtesy of Eleanor Harvey.* Each (VG-NM) : $25-35.

Rich Bay 9" Arabian Stallion, Tenite, Paola sample pictured in her 1987 catalog (as #421), but not put into production; it was replaced by Chestnut, fewer than five were made. *Courtesy of Sandy Tomezik.*

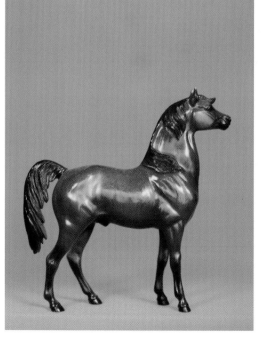

Dark Bay (Pearled) 9" Arabian Stallion with no white; styrene, one-of-a-kind, a 1984 sample sold in 1994.

9" Arabian Stallions in 1960s Sorrel (Tenite), Glossy Chestnut from Black Horse Ranch (styrene), and 1980s Sorrel, which is glossy and styrene. The 1960s model has *eye* whites. Not shown: Light Sorrel (brown body, white mane and tail), styrene, May 1987 Paola test color #337A, 35 made. VG-NM: $30-55 for #337A only.

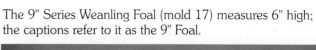

The 9" Series Weanling Foal (mold 17) measures 6" high; the captions refer to it as the 9" Foal.

Palomino 9" Foal, Tenite, Hartland Plastics #6100 series, 1964-1968. $8-13.

Sorrel 9" Foal, Tenite, Hartland Plastics #6100 series, 1964-1968. $8-13.

Bay: Blood Bay 9" Foal, Tenite, Hartland Plastics #6100 series, 1964-1968. $8-13.

Light Bay 9" Foal, Tenite, Paola #491, 1988-1990. VG-NM: $20-30.

Chestnut 9" Foal, Tenite, Paola #492, 1988-1990. VG-NM: $20-30.

"Bay Roan" 9" Foal, Tenite, a Steven special of 400-500 for the 1994 West Coast Model Horse Collector's Jamboree. VG-NM: $25-40.

Light Bay Appaloosa 9" Foal, Tenite, Paola #494, 1988-1990. *Courtesy of Traci Durrell-Khalife*. VG-NM: $20-35.

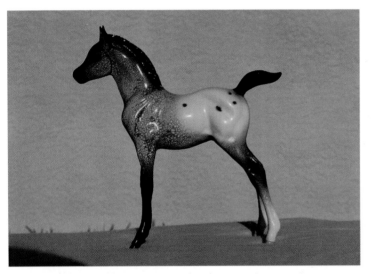

Red Roan Appaloosa 9" Foal, Tenite, Paola #493, 1988-1990. *Courtesy of Carla S. Clifford*. VG-NM: $20-35.

Dapple Grey 9" Foal, Tenite, Paola #490, 1988-1990. *Courtesy of Peggy Howard*. VG-NM: $20-30.

Metallic Blue Roan 9" Foal, Tenite, Paola #495, 1989-1990. VG-NM: $25-40.

Black 9" Foal with silver mane and tail and blue undertone, Tenite, Paola, 30 made, painted in 1990, sold in 12/1994. VG-NM: $25-35.

Black 9" Foal, Tenite, Paola, "Black Embers," a special run of 250 for Black Horse Ranch, 1990; it has a (white) star on its face and a left hind sock and pale hoof under it. VG-NM: $25-35.

Glossy Black 9" Foal, styrene, Steven (sold by Paola), only 10 or 11 made, Christmas 1985. *Courtesy of Shirley Ketchuck.* Not shown: Glossy Dark Bay (called "Chestnut" on the flyer) styrene, by Steven, Christmas 1985, 10 or 11 made. VG-NM: $30-40 each.

Pearl White 9" Foal, Tenite, Paola #490P, special run color, 1988. *Courtesy of Fleanor & Shay Goosens.* VG-NM: $25-35.

The Pearl White 9" Foal (Paola #490P) can look iridescent in certain lighting conditions. *Courtesy of Traci Durrell-Khalife.* $25-35.

Pearl White 9" Foal with pink hooves; Tenite, a 1990 sample color by Paola, only two made in this color. *Courtesy of Eleanor Harvey.*

Amid the rosettes, this metallic gold bay 9" Foal in Tenite by Paola Groeber was a prize in St. Louis, 1990. *Courtesy of Paola Groeber.*

Paola donated this metallic gold 9" Foal with white points to a model show in St. Louis, July 1990. *Courtesy of Paola Groeber.*

Gold 9" Foal with high white points and much white on face; Tenite, a 1990 sample color by Paola: six were made. *Courtesy of Eleanor Harvey.*

This 9" Foal in blue with white points (Tenite, by Paola) was a prize in St. Louis, 1990. *Courtesy of Paola Groeber.*

Wedgewood (Blue) Foal with high white points and much white on face; Tenite; a 1990 sample color by Paola: 12 were made. *Courtesy of Eleanor Harvey.* VG-NM: $35-65+.

Copper 9" Foal with white mane and tail, Tenite, a Paola special of six pieces, signed and numbered, c.1990; pink front hooves. *Courtesy of Shirley Ketchuck.*

The 9" Series Mustang with smooth, not woodcut, surface (mold 18) is 9" to 9.25" high.

Far left: Sooty Dun Mustang, Tenite, Steven #220-20, 1993 special for West Coast Model Horse Collector's Jamboree, 400 made. *Model, courtesy of Sande Schneider*. VG-NM: $30-60.

Left: Dapple Grey Mustang (with white mane and tail), styrene, Steven #242, 1985-1986. $10-20.

Right: Dark Dapple Grey Mustang (with black points), Tenite, Paola #216S, a 1989 special of 100; some were re-sold by Your Horse Source, 1990, as #H215/H216. *Model, courtesy of Sande Schneider*. VG-NM: $35-50.

Far right: White-Grey Mustang, Tenite, Hartland Plastics #8001 Classic series, 1967. *Model, courtesy of Sande Schneider*. $15-35.

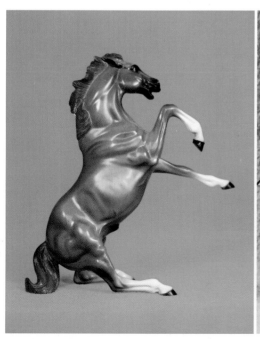

Far left: Red Dun (Chestnut) Mustang, Tenite, Steven #247, "Flame's Dawn," 1988-1990. $12-25.

Left: Red Dun Mustang, Tenite, Paola #440, 1988-1990. VG-NM: $25-40.

Right: Palomino Mustang, Tenite, Hartland Plastics #889, 1965-1966. $15-35.

Far right: Metallic Gold Buckskin Mustang with black points (smooth, not woodcut), Tenite, Hartland Plastics #8679, 1967. *Courtesy of Daphne R. Macpherson.* $15-40.

Far left: Red Bay Mustang, Tenite, Steven #220, "Desperado," 1993-1994. $12-30.

Left: Black Mustang, Tenite, Paola #215S, a 1989 special of 100 made; some were re-sold by Your Horse Source, 1990, as #H216/#H215. *Model, courtesy of Sande Schneider.* VG-NM: $35-50.

Right: 1960s Black Pinto Mustang, Tenite, Hartland Plastics #885, 1964-1966. $12-25.

Far Right: Left side of 1960s, Tenite, Black Pinto Mustang, Hartland Plastics #885, 1964-1966.

Far left: The Black Pinto Mustang by Steven, #238, from 1984-1986 (in styrene) lacks the right hip spot of the 1960s model. *Model, courtesy of Cecile Bellmer.* $10-22.

Left: The left side of the Steven #238 styrene Black Pinto Mustang, 1984-1986, lacks the shoulder spot of the 1960s model. *Model, courtesy of Cecile Bellmer.*

Right: Some 1984 examples of the Steven #238, styrene Black Pinto Mustang are missing from their right side both the hip and neck spots of the 1960s model. *Courtesy of Jan Kreischer.* $12-25.

Far right: The early pieces of the Steven #238 styrene 1984-1986 Black Pinto that are missing two spots on the right side have a full set of three spots on the left side so that the left side matches the 1960s model. *Courtesy of Jan Kreischer.*

Far left: Brown Pinto Mustang, Tenite, Steven #248 and Paola #441, 1988-1990. *Model, courtesy of Sande Schneider.* Steven: $12-25; Paola (VG-NM): $25-35.

Left: Left side of Brown Pinto, Tenite Mustang, Steven #248 and Paola #441. *Model, courtesy of Sande Schneider.*

Brown Pinto Mustang in styrene (Glossy), one-third of the Steven #278 Horse Assortment, "Three-Stallion" set for the 1991 JCPenney Christmas catalog; up to 20,000 sets were made. $10-22.

Left side of styrene Brown Pinto Mustang by Steven in styrene for the 1991 JCPenney catalog.

The Blue Roan Pinto Mustang, Tenite, Paola #442, 1990 only. The production model (*right*) has eye whites; a test model (*left*) has no eye whites and is a deeper blue. *Models, courtesy of Donna Anderson; photo, courtesy of Ellen W. Vogel.* Production model (VG-NM): $35-60.

Buckskin Pinto Mustang, Tenite, a c.1990 Paola test color sold in 1994; seven were made. *Courtesy of Karen Oelkers.* VG-NM: $40-75.

Left sides of the two Blue Roan Pinto Mustangs, Paola #442, 1990. *Models, courtesy of Donna Anderson; photo, courtesy of Ellen W. Vogel.*

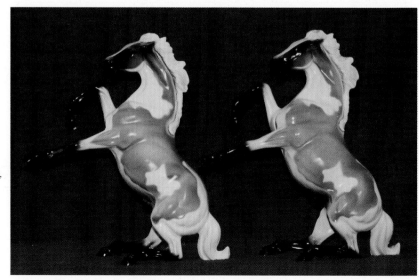

The Hong Kong copy in dark gray (*left*) is 8.5" high, compared to 9.25" for the 1980s Steven Hartland Mustang in dapple grey (*right*). Both are styrene. The Hartland is marked, "Hartland." Copy: $4-9.

Above left: Hong Kong copies of the Hartland Mustang include this 8.5" high bay marked "P in a wide diamond/654/Made in Hong Hong." It is styrene. Other copies were sold with lamps in the 1980s. Copies: $4-9.

Right: Silver Mustang with black points, a Steven model intended as a special for Cascade Models/1994 Northwest Congress model horse show, but was never produced; only two were made. *Courtesy of Daphne R. Macpherson.*

Far right: Silver Mustang with white points, Tenite, only one made, by Paola, 1990, for a model show in St. Louis. *Courtesy of Paola Groeber.*

Metallic Gold Pinto Mustang, Tenite; unique; Paola donated it to a model show in St. Louis, July 1990. *Courtesy of Paola Groeber.*

Far left: Taffy Sorrel or Deep Palomino 9" Mustang with white points, styrene, c.1986; the first model Paola Groeber ever painted; it is unique. *Courtesy of Tina English-Wendt.*

Left: Buckskin 9" Mustang, Tenite, by Paola, 1990; fewer than 10 made, possibly unique. *Courtesy of Eleanor Harvey.*

The Woodcut Mustang, with its whittled-looking surface (mold 19) measures 9.25" high and is part of the 9" series.

Four of the 9" series Mustangs: Steven #242 Dapple Grey, 1960s palomino, Steven #247 Red Dun (chestnut), and Blue Roan Pinto (Paola #442).

Walnut Mustang, Tenite, Hartland Plastics #886W, 1964-1968. $15-30.

Right: Cherry Mustang, Tenite, Hartland Plastics #886C, 1964-1966. *Model, courtesy of Sande Schneider.* $15-30.

Far left: Ebony Mustang, Tenite, Hartland Plastics #886E, 1964-1966. $18-35.

Shiny Walnut Mustang, Durant Plastics #4202, 1970-1973. *Model, courtesy of Jackie Himes; photo, courtesy of April and Jon Powell.* $15-25.

All-Gold Mustang (woodcut), produced by Hartland Plastics for beer signs. Note the two holes drilled in the barrel and stifle where the model attached to the wall sign. Models with holes in the feet went with a counter display. $30-60.

Left side of All-Gold Mustang used only with beer signs. This color was never in the Hartland model line. The 1960s, metallic gold Mustang with black points was from the smooth, not woodcut, Mustang mold.

The 9" Series Tennessee Walker (mold 20), a woodcut model, measures 8" high and came in two mold variations: models with the mane on the left side of the neck, and models with the mane equally long on both sides.

Walnut 9" Tennessee Walker with mane on left, Tenite, Hartland Plastics #887W, 1964-1965. $18-30.

Left side of 9" Walnut Tennessee Walker with mane on left.

Cherry 9" Tennessee Walker with mane on left, Tenite, Hartland Plastics #887C, 1964-1965. $20-35.

Shiny Cherry 9" Tennessee Walker with mane on left, Tenite, Hartland Plastics #887C, 1964-1965; many had some black stain added, but this one has none, so the color of the red-brown plastic is evident. *Model, courtesy of Jaci Bowman.* $20-35.

Ebony 9" Tennessee Walker with mane on left, Tenite, Hartland Plastics #887E, 1964-1965. This model photographed as plain black although it has highlights (stain) in the crevices. *Courtesy of April and Jon Powell.* $20-35.

This photo captures the grayish-blue highlighting (stain) Hartland Plastics typically applied to its Ebony 9" Tennessee Walkers, #887E. With mane on left, this model dates to 1964-1965. *Model, courtesy of Terry Davis.* $20-35.

Walnut 9" Tennessee Walker with two-sided mane, Tenite, Hartland Plastics #887W, 1965-1968. $20-35.

Left side of Walnut 9" Tennessee Walker with two-sided mane.

Cherry 9" Tennessee Walker with two-sided mane; Tenite, Hartland Plastics #887C, 1965-1966. *Model, courtesy of Sande Schneider.* $25-40.

Left side of Cherry 9" Tennessee Walker with two-sided mane. *Model, courtesy of Sande Schneider.*

The 9" series, Three-Gaited Saddlebred (mold 21), is a woodcut measuring 8.5" high.

Ebony 9" Tennessee Walker with mane on both sides, Tenite, Hartland Plastics #887E, 1965-1966. *Courtesy of April and Jon Powell.* $25-40.

Walnut Woodcut 9" Three-Gaiter with stain and smooth, but not highly shiny, finish; Tenite, Hartland Plastics #888W, 1965-1968. $20-35.

Cherry Woodcut 9" Three-Gaiter; Tenite, Hartland Plastics #888C, 1965-1966. $20-40.

Ebony Woodcut 9" Three-Gaiter; Tenite, Hartland Plastics #888E, 1965-1966. $20-40.

Ebony, Walnut, and Cherry 9" Three-Gaiters by Hartland Plastics, 1960s.

Shiny Walnut 9" Three-Gaiter (with little stain); Tenite, Durant Plastics #4202, 1970-1973. $18-35.

This metal horse with metal reins and saddle resembles Hartland's 9" Three-Gaiter. It is about 6" high. *Model, courtesy of Stephanie M. Jones, who bought it in an antique store in Iowa.*

The stallion from the woodcut 7" Saddlebred family is similar to the 9" woodcut Saddlebred in every way but size.

The 11" Series Arabians (mold 22)
measure 9.5 to 9.75 inches high, and are
referred to as "11 inch."

Matte, Red Bay 11" Arabian (with white face and stockings), Tenite, Hartland Plastics #99131, 1968. *Courtesy of Susan Bensema Young.* AV: $20...VG: $35...EX: $60...NM: $85.

Glossy, Red Bay 11" Arabian (with white stockings and eye whites), Tenite, Hartland Plastics #9916, 1967. *Courtesy of Susan Bensema Young.* AV: $20...VG: $40...EX: $75...NM: $100.

In this photo, the Glossy Red Bay 11" Arabian (in Tenite, Hartland Plastics #9916, 1967) looks iridescent or lustrous. *Courtesy of Peggy Howard.* AV: $20...VG: $40...EX: $75...NM: $100.

Light Bay 11" Arabian, styrene, Paola #223A, May 1987 test color, 50 made. *Courtesy of Sandy Tomezik.* VG-NM: $25-$40.

Dusty Bay 11" Arabian (with black body shadings and black mane, tail, and lower legs), Tenite, Hartland Plastics #99141, 1968. *Courtesy of Tina English-Wendt.* AV: $20...VG: $35...EX: $60...NM: $85.

Tan Bay 11" Arabian (with no black shadings on the body, but black mane, tail, and lower legs), Tenite, Durant #4500, 1970-1973. *Courtesy of Tina English-Wendt.* AV: $15...VG: $30...EX: $50...NM: $75.

Chestnut 11" Arabian with darker brown mane, tail, and hooves (some reportedly had black m-t-h, in which case they would be bay), styrene, Steven #205 "Light Bay," 1983-1986. $15-$40.

Sorrel 11" Arabian with flaxen mane and tail, Tenite, Hartland Plastics #90021, 1968. AV: $20...VG: $35...EX: $60...NM: $85.

Glossy Palomino 11" Arabian, Tenite, Hartland Plastics #9914, 1967. *Model, courtesy of Sande Schneider.* AV: $20...VG: $35...EX: $60...NM: $85.

Matte Palomino 11" Arabian, Tenite, Hartland Plastics #9914, 1967. *Courtesy of Eleanor Harvey.* AV: $20...VG: $35...EX: $60...NM: $85.

Rose Grey 11" Arabian, Tenite, Paola #390 and Steven series #262, both 1988-1990. *Courtesy of Eleanor Harvey.* Steven: $15-40; Paola (VG-NM); $25-$50.

Pearl Blue 11" Arabian (with white mane, tail, lower legs, and hooves), styrene, Steven #209, 1985-1986. $20-$45.

Charcoal Gray 11" Arabian (with matte finish and black m-t-ll-h),Tenite, Hartland Plastics #99121, 1968. *Courtesy of Eleanor Harvey.* AV: $20...VG: $35...EX: $60...NM: $85.

Flea Bit Grey 11" Arabian, Tenite, Steven #204 "Royal Elite," 1993-1994. $20-$45. Not shown: Dark Dapple Grey (styrene), Paola #223B, May 1987, 70 made. VG-NM: $30-45 for #223B.

White Grey 11" Arabian Stallion (with gray mane and tail), Tenite, Hartland Plastics #9912, 1967. AV: $20...VG: $35...EX: $60...NM: $85.

Pearl White 11" Arabian Stallion (painted over golden-yellow plastic), Tenite, Hartland Plastics #9915, 1967. *Courtesy of Anni Stapley-Koziol.* AV: $20...VG: $40...EX: $75...NM: $100; can go higher at auction.

Pearl White 11" Arabian Stallion (painted over white plastic), Tenite, Hartland Plastics #9915, 1967. *Model, courtesy of Jackie Himes; photo, courtesy of April and Jon Powell.* AV: $20...VG: $40...EX: $75...NM: $100; can go higher at auction.

Pearl White 11" Arabian Stallion, Tenite, by Paola, only one or two were made in this color, 1989. *Courtesy of Eleanor Harvey.*

Jet Black 11" Arabian with silver pearl mane, tail, and stockings, Tenite, Hartland Plastics #9913, 1967. *Model, courtesy of Sande Schneider.* AV: $20...VG: $40...EX: $75...NM: $100; can go dramatically higher at auction.

Raven Black 11" Arabian, styrene, Steven #204, 1983-1986. $15-$35.

The 11" Series Five-Gaited Saddlebred (mold 23)
measures 9.5-10" high, and is called "11 inch."

Chestnut 11" Saddlebred with darker brown mane, tail, hooves, and pasterns; 1983-1986, Steven # 202 "Light Bay," styrene. *Model, courtesy of Jackie Himes; photo, courtesy of April and Jon Powell.* $15-$35.

Dark Bay 11" Saddlebred with white face and stockings; Tenite, Hartland Plastics #99331, 1968. *Model, courtesy of Jackie Himes; photo, courtesy of April and Jon Powell.* AV: $20...VG: $40...EX: $75...NM: $100.

Bay 11" Saddlebred with black legs; Tenite, Hartland Plastics #9933, 1967. *Courtesy of Eleanor Harvey.* AV: $20...VG: $40...EX: $75...NM: $100.

Bay 11" Saddlebred with white stockings; Tenite, Hartland Plastics #9937, 1967. *Courtesy of Eleanor Harvey.* AV: $20...VG: $40...EX: $75...NM: $100.

Copper Sorrel 11" Saddlebred, (with white mane, tail, hooves, and lower legs); Tenite, Hartland Plastics #9935, 1967. *Courtesy of Eleanor Harvey.* AV: $20...VG: $40...EX: $75...NM: $100.

Dusty Bay 11" Saddlebred (with white stockings), Tenite, Hartland Plastics #99321, 1968. *Courtesy of Eleanor Harvey.* AV: $20...VG: $35...EX: $60...NM: $85.

Liver Chestnut 11" Saddlebred, styrene, May 1987, Paola test color #220A, 24 made. *Courtesy of Sandy Tomezik.* VG-NM: $30-$45.

Light Sorrel 11" Saddlebred (with white mane and tail), styrene, May 1987, Paola test color #220B "Light Chestnut," 31 made. VG-NM: $30-$45.

Orange Bay 11" Saddlebred, Tenite, Hartland Plastics #99351, 1968 and Durant #4502, 1970-1973. With a paint run on the left foreleg and over spray at the left hind hoof, this example is probably by Durant. *Courtesy of Eleanor Harvey.* AV: $15...VG: $30...EX: $45...NM: $60; less for untidy paint.

Red Bay 11" Saddlebred (with white face and stockings), Tenite, Hartland Plastics #9936 Mahogany, 1967. AV: $15...VG: $30...EX: $50...NM: $75.

This Red Bay, 11" Saddlebred is unusual because it has no dark body shadings; Tenite, Hartland Plastics #9936, 1967. *Courtesy of Sandy Tomezik.* AV: $15...VG: $30...EX: $50...NM: $75.

Palomino 11" Saddlebred, styrene, Steven #208, 1985-1986. $20-$45.

11" Saddlebred, Buckskin, Tenite, Hartland Plastics, 1967 or 1968 (not in catalogs, but common). AV: $20...VG: $35...EX: $60...NM: $85.

"Dapple Rose Grey" 11" Saddlebred with dark points, Tenite, Steven #202, 1993-1994. $20-45.

Charcoal Gray 11" Saddlebred, Tenite, Paola # 395 and Steven series #262, 1988-1990. *Courtesy of Eleanor Harvey*. Steven: $15-40; Paola (VG-NM): $25-50.

White 11" Saddlebred with black mane and tail; Tenite, Hartland Plastics #9922, 1967. *Courtesy of Eleanor Harvey*. AV: $20...VG: $40...EX: $75...NM: $100.

White 11" Saddlebred with gray mane and tail; Tenite, Hartland Plastics #9922, 1967. *Model, courtesy of Sande Schneider*. AV: $20...VG: $40...EX: $75...NM: $100.

Black 11" Saddlebred with white socks, styrene, Steven #203 Raven Black, 1983-1986. $15-$40.

Cascade Models added a blaze to 33 of the Steven #201-05 Blue-Black 11" Saddlebreds for Shay Goosens' model horse show in Virginia, 1994. Some were sold then and others were offered at her show in April 2000. *Courtesy of Eleanor & Shay Goosens*. Fair (peeling paint) to excellent: $20-45.

Black 11" Saddlebred with black legs; Tenite, Steven #201-05 Blue-Black, a spring 1993 special for Cascade Models, 150-200 made, acetate. *Courtesy of Eleanor Harvey*. Fair (peeling paint) to excellent: $20-$45.

Black Pinto 11" Saddlebred, Tenite, a 1993 Steven special raffled at West Coast Model Horse Collector's Jamboree; 20 were made, all with different pinto patterns. *Courtesy of Eleanor Harvey.* VG-NM: $40-$100; can go dramatically higher at auction.

Another example of the Black Pinto 11" Saddlebred, the Tenite, Steven (no #), special raffled at West Coast Model Horse Collector's Jamboree, 1993. *Courtesy of Daphne R. Macpherson, Cascade Models.* VG-NM: $40-$100 and up.

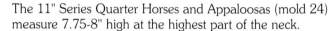

The 11" Series Quarter Horses and Appaloosas (mold 24) measure 7.75-8" high at the highest part of the neck.

Charcoal Gray 11" Saddlebred with white mane and tail, a one-of-a-kind model by Paola Groeber, c. 1990. *Courtesy of Eleanor Harvey.*

Dapple Sooty Buckskin 11" Quarter Horse; Tenite, Steven #200 "Prairie Dancer," 1993-1994. $20-$45.

Claybank Bay 11" Quarter Horse, in matte finish with black mane, tail, hooves, and lower legs; styrene, Steven #200 Light Bay, 1983-1986. $15-40.

Claybank Bay 11" Quarter Horse, in glossy finish with black mane and tail, but dark brown hooves and lower legs; styrene, Steven #200, 1983-1986. *Courtesy of Maggi Jacques.* $15-$40.

Sorrel (Pearled) 11" Quarter Horse (with flaxen mane and tail); Tenite, Paola # 400S, special run color, 1989. *Courtesy of Eleanor Harvey.* VG-NM: $40-$85.

Red Roan 11" Quarter Horse; Tenite, Paola #401, a special sold with a hand tooled, leather saddle, bridle, etc.; 50 made, 1988. *Courtesy of Eleanor Harvey.* VG-NM: $100-$175 with saddlery; $40-$95 without saddlery.

The Red Roan 11" Quarter Horse (Paola #401) was supposed to have two hind stockings, but this model has none (and it's not the red dun color, either). *Courtesy of Judith Miller.* VG-NM: $40-85.

Two 1960s, Bay 11" Quarter Horses with white stockings: Regal Bay #99231 from 1968 (*left*), and Superb Bay #9924 from 1967. The Superb Bay has richer color. Both are Tenite, both by Hartland Plastics. *Courtesy of Sandy Tomezik.* AV: $20...VG: $45...EX: $80...NM: $125; can go higher at auction.

Superb Bay 11" Quarter Horse, Tenite, Hartland Plastics #9924, 1967. *Courtesy of Susan Bensema Young.* AV: $20...VG: $45...EX: $80...NM: $125; can go higher at auction.

This bay 11" Quarter Horse without body shadings was painted reddish-brown (with black lower legs) over golden yellow plastic. It's probably a variation of #9924 or #9923. *Courtesy of Sandy Tomezik.* AV: $20...VG: $45...EX: $80...NM: $125.

Maroon Bay 11" Quarter Horse (with black legs), Tenite, Hartland Plastics #9923, 1967. *Courtesy of Tina English-Wendt.* AV: $20...VG: $45...EX: $80...NM: $125.

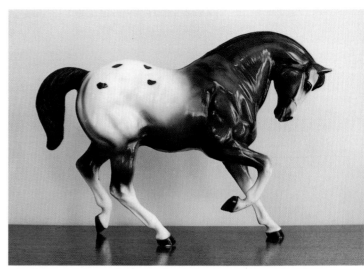

Red Bay 11" Appaloosa (with white stockings), Tenite, Hartland Plastics #9926, 1967. *Courtesy of Eleanor Harvey.* AV: $20...VG: $45...EX: $80...NM: $125; can go dramatically higher at auction.

This Red Bay 11" Appaloosa (Tenite, #9926, 1967) has black socks below white stockings. The catalog showed the model that way, but most just had white stockings. *Courtesy of Jan Kreischer.* AV: $20...VG: $40...EX: $75...NM: $100.

11" Light Brown Appaloosa, an early version of the Steven #201 Appaloosa, 1983-1986. Its mane, tail, hooves, and lower legs are very dark brown, but not quite black. *Model, courtesy of Sande Schneider.* $20-$50.

11" Brown Appaloosa with glossy finish, reddish brown body, and darker brown mane and tail; another version of Steven #201 Appaloosa, 1983-1986. $15-$40.

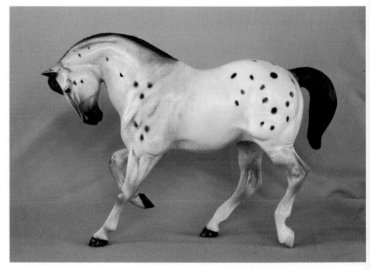

11" Red Roan Leopard Appaloosa, Tenite, Paola #403, 1988-1990. *Courtesy of Traci Durrell-Khalife.* VG-NM: $30-$75.

11" Black Appaloosa with less apparent mottling where black areas meet white; Tenite, Paola #400, 1988-1990. *Courtesy of Susan Bensema Young.* VG-NM: $30-$75.

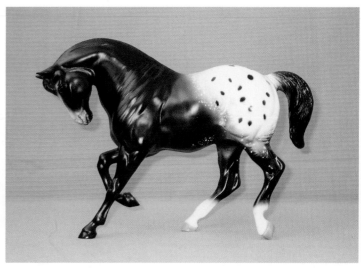

11" Black Appaloosa with mottling more apparent, Tenite, Paola #400, 1988-1990. *Courtesy of Peggy Howard.* VG-NM: $30-$75.

11" Dark Blue Appaloosa, Tenite, Hartland Plastics #99271, 1968. *Courtesy of Peggy Howard.* AV: $20...VG: $45...EX: $80...NM: $125; can go dramatically higher at auction.

Metallic Blue 11" Quarter Horse with white points; Tenite, Hartland Plastics #9927, 1967. *Courtesy of Eleanor Harvey.* $20...VG: $45...EX: $80...NM: $125; can go dramatically higher at auction.

Palomino 11" Quarter Horse, Tenite, Paola #402 and Steven series #262, 1988-1990. *Courtesy of Eleanor Harvey.* Steven (AV-EX): $25-50; Paola (VG-NM): $35-75.

A unique 11" palomino Quarter Horse by Paola (Tenite, c.1990) with tan hooves and shadings; it was owned by Marney Walerius, who called it "Dublin Gold." *Courtesy of Eleanor Harvey.*

Red Dun 11" Quarter Horse, Tenite, "Dun in Dreams," by Paola, a special for Black Horse Ranch, 250 made, 1990. They are numbered. VG-NM: $35-75.

Dark ("Golden Tan") Buckskin 11" Quarter Horse, Tenite, Durant #4501 (1970-1973). *Courtesy of Eleanor Harvey.* AV: $20...VG: $35...EX: $50...NM: $75.

Light ("Honey") Buckskin 11" Quarter Horse, Tenite, Hartland Plastics #99221 (1968). *Courtesy of Eleanor Harvey.* AV: $20...VG: $35...EX: $50...NM: $75.

Creamy Dun (Creamy Buckskin) 11" Quarter Horse, Tenite, Paola #404, 1989-1990. *Courtesy of Susan Bensema Young.* VG-NM: $30-60.

Dark Dapple Grey 11" Quarter Horse, Tenite, by Paola, a special run for Marney Walerius' Model Horse Congress in northern Illinois, 50 made, 1989. *Courtesy of Eleanor Harvey.* VG-NM: $35-85.

"Forest Dew," Dark Dapple Grey 11" Quarter Horse, Tenite, Steven #200-20, special for Modell Pferde Versand, Germany; 200 made, 1994. *Courtesy of Eleanor Harvey.* VG-NM: $35-75.

Light Dapple Grey 11" Quarter Horse, Tenite, by Paola, a special for two model horse shows (Model Horse Congress, Illinois (models dated 6/89), and Heather Wells' show, California); about 88 made all together, 1989. *Courtesy of Laura Pervier.* VG-NM: $35-$85.

Grulla (Pearled) 11" Quarter Horse, Tenite, Paola #400P, special run color, 1989. VG-NM: $35-$85.

Metallic Silver 11" Quarter Horse with black m-t-h; by Paola, a special color, 12 pieces numbered between 1 and 30; painted c. 1990 and sold in December 1994. *Courtesy of Judith Miller.* VG-NM: $35-$85.

Pearl White 11" Quarter Horse with two black feet; Tenite; a Paola special of 12 painted in 1991 and sold in 1994; they are signed and dated. VG-NM: $50-100. *Courtesy of Eleanor Harvey.*

Some examples of the special of 12 Pearl White 11" Quarter Horses sold by Paola in 1994 have four pink feet, instead of two pink and two black feet. *Courtesy of Susan Bensema Young.* VG-NM: $50-100.

Albino 11" Quarter Horse (pale, blue-white body with pink areas), variation with black hooves; styrene, Steven #207, 1985-1986. *Courtesy of Maggi Jacques.* $15-$40.

Albino 11" Quarter Horse (pale, blue-white body with pink areas), early variation with pale hooves; styrene, Steven #207, 1985-1986. $15-$40.

1960s White-Grey 11" Quarter Horse (with gray mane and tail, but no body shadings), Tenite, Hartland Plastics #9922, 1967. *Courtesy of Tina English-Wendt.* AV: $20...VG: $45...EX: $80...NM: $125; can go dramatically higher at auction.

White-Grey Jamboree 11" Quarter Horse (with gray body shadings), Tenite, Steven (no #), special for West Coast Model Horse Collector's Jamboree raffle, etc., 25 made, 1992. *Courtesy of Judith Miller.* VG-NM: $35-$85.

Copenhagen (Blue Dapple) 11" Quarter Horse, by Paola, one made; it was a prize at a model horse show in St. Louis, 1990. *Courtesy of Kelly (Engelsiepen) Scotti.*

It's interesting that three legs on this 10.25" black, toy horse are posed like the Hartland 11" Quarter Horse. The brand name is not marked. It is, of course, not a Hartland horse. $3-8.

Midnight Blue 11" Quarter Horse with silver base coat, Tenite, Paola, c.1990, three made. *Courtesy of Shirley Ketchuck.*

Sooty Buckskin 11" Quarter Horse with silver base coat, Tenite, Paola, c.1990, seven made. *Courtesy of Shirley Ketchuck.*

Red Bay 11" Quarter Horse with silver base coat, Tenite, Paola, five this color, 1989. *Courtesy of Eleanor Harvey.*

Metallic Gold 11" Quarter Horse (with black mane and tail); Paola, c. 1990, one made. *Courtesy of Shirley Ketchuck.*

Metallic Copper 11" Quarter Horse (with white mane and tail), Tenite, by Paola, three made; one raffled at Daphne Macpherson's Hartland model horse show, Washington state, November 1990; another was a prize in St. Louis, 1990. *Courtesy of Daphne R. Macpherson.*

The 11" Series Grazing Mares (mold 25) measures 6.5" high at the highest
point—the withers—and is referred to as "11 inch."

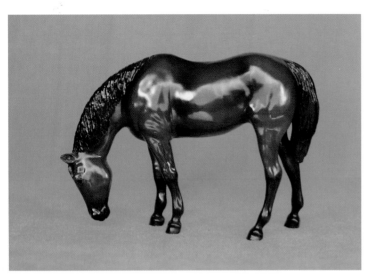

Dark Bay 11" Grazing Mare, styrene, Steven #206, 1985-1986. $8-$18.

The 11" series Hartland Grazing Mare is very similar (in all but the head
and neck position) to the Quarter Horse Yearling by Breyer Animal Creations. Steven Mfg. purchased the 11" series Grazing Mare design and
tooling in China in the 1980s and molded it in Missouri; thus, it became
part of the Hartland horse line.

Black Appaloosa 11" Grazing Mare; Tenite, Paola #385 "Grazing POA
(Pony of the Americas)," and Steven, series #262, 1988-1990. Steven:
$15-$30; Paola (VG-NM): $25-35.

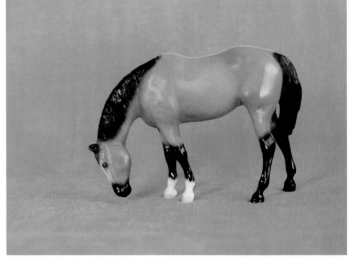

Bright Buckskin 11" Grazing Mare; Tenite, Steven #206, "Gold Duster,"
1993-1994. $15-$35.

The Lady Jewel and Jade (Arabian Mare and Foal) models are scaled to the 11" series. Lady Jewel ("Jewel" for short) is 8" high at the highest part of her mane; the Jade mold is 6.25" high. They are molds 26 & 27.

Left: Black Jade, Tenite, in Steven set #233 with dapple grey mare; foal's name is "Blessing," 1993-1994. VG-NM: $15-$25.

Far left: Dapple Grey Lady Jewel, Tenite, Steven #233 "Bedouin Princess" in set with black foal, "Blessing," 1993-1994. VG-NM: $30-$50.

Liver Chestnut Pinto Jewel and Jade (mostly brown), Tenite, Steven set #233-20, "Gloria and Nimbus," a 1994 special for Modell Pferde Versand, Klettgau, Germany. *Courtesy of Ingrid Muensterer.* Average condition (paint problems)—Mare: $15-$25; Foal: $10-$15; could go higher at auction.

Left side of "Gloria and Nimbus," Steven #233-20, 1994 special for Modell Pferde Versand; the 300 sets were recalled due to defective paint, but some purchasers kept them. *Courtesy of Ingrid Muensterer.*

Chestnut Pinto Lady Jewel (mostly white), Tenite, a 1992 Steven special for Black Horse Ranch; shapes and locations of brown areas vary; 333 sets with matching foal. *Model, courtesy of Sande Schneider.* VG-NM: $40-65;

Left side of Chestnut Pinto Lady Jewel, the 1992 Steven special for Black Horse Ranch. *Model, courtesy of Sande Schneider.*

Chestnut Pinto Jade (mostly white), a 1992 Steven (no #) special for Black Horse Ranch; shapes and locations of brown areas vary; 333 sets with matching mare. *Model, courtesy of Sande Schneider*. VG-NM: $20-$30.

Left side of Chestnut Pinto Jade, the Steven 1992 special for Black Horse Ranch. *Model, courtesy of Sande Schneider*.

Black Appaloosa Jade, Tenite, Steven (no #), "Black Blanket Appaloosa," special for Black Horse Ranch, 1992; sold in set with matching mare; 333 sets made. *Model, courtesy of Sande Schneider*. VG-NM: $25-35.

Black Appaloosa Lady Jewel, Tenite, Steven (no #), "Black Blanket Appaloosa," special for Black Horse Ranch, 1992; 333 sets with matching foal. *Model, courtesy of Sande Schneider*. VG-NM: $40-65.

Black Lady Jewel (with silver mane and tail), Tenite; a 1989, a Paola show special, sold with matching foal. *Model, courtesy of Sande Schneider*. VG-NM: $60-100.

Black Jade with silver mane and tail, Tenite; a 1989, a Paola show special, sold with matching mare. *Model, courtesy of Sande Schneider*. VG-NM: $30-50.

Bay Jade with body color molded in; Tenite, from Steven set #256-14, "Lady Jewel & Jade," special for 1992 JCPenney Christmas catalog; sold with bay Lady Jewel. $10-$25.

Bay Jade with body color painted on; Tenite, from Steven set #256-14, a special for 1992 JCPenney Christmas catalog; sold with bay Lady Jewel. *Courtesy of Sandy Tomezik*. $10-$25.

Bay Lady Jewel, Tenite; from Steven set #265-14, "Lady Jewel & Jade," a special for the JCPenney 1992 Christmas catalog. $15-$40.

Rose Grey Lady Jewel, Tenite, Paola #700; goes with chestnut foal, but was sold separately; 1988-1990. VG-NM: $60-$100.

Left: Chestnut Jade (red chestnut), Tenite, Paola #710, goes with rose grey mare, but was sold separately; 1988-1990. $30-50.

Right: The Arabian Foal in the Best Talking Horses series (styrene, 6" high, 1997 by Best Card Company, Inc., Douglasville and Austell, Georgia) is sometimes compared to Hartland's Jade. The Best horses were made in China. The foal came in black and in golden chestnut. $8-12.

The Best Talking Horses Arabian Mare (styrene, 7.75" high, 1997 by Best Card Company, Inc.) may have been inspired by Hartland's Lady Jewel. The "golden chestnut" Arabian mare and foal were Best set #302; the Best Arabian mare also was painted in a "flea-bitten gray," which came with a black foal. $12-18.

Bright Sorrel Lady Jewel and Jade; Tenite, Steven set #941, "Selyyn and Soleil," a special for The Equine Center, 1994, 300 sets, of which many had defective paint. *Courtesy of Peggy Howard.* With typical paint damage-Mare: $25; Foal: $15; could go higher at auction. Sets with perfect paint are rare and worth dramatically more.

"Kerry's Jubilee," the cantering, bright chestnut Arabian stallion, makes a nice threesome with Hartland's Lady Jewel and Jade. The stallion is part of the Winner's Choice, "Gold Standard" series by Creata International Ltd., Hong Kong. The series, by designer/sculptor Candace Liddy, debuted in 1998. Mass market colors, AV-EX: $5-15; specialty colors, including "Kerry's Jubilee," VG-NM: $15-25.

Six horse sculptures cast in resin were painted by Paola Groeber and sold under the Hartland name as limited editions in the late 1980s-early 1990s. The sculptors were Carol Gasper, Linda Lima, and Kathleen Moody. Since these models are resin, they weigh more than plastic models. They are molds R1-R6.

Arabian Gelding resin (mold R5) in Dapple Grey, Paola #890, "Shariff," a limited edition of about 38, 1990-1991; sculpture by Linda Lima. *Model, courtesy of Bobbie Mosimann; photo, courtesy of Michelle Grant.* NM: $200.

Arabian Stallion resin (mold R6), "Simply Splendid," in Dapple Grey; sculpture by Kathleen Moody, painted by Paola Groeber. Paola painted 12 this color; about 88 more in various colors were sold by Kathleen Moody or by Paola. *Courtesy of Shirley Ketchuck.* NM: $200.

Blue Roan Peruvian Paso (resin, mold R2); sculpture by Kathleen Moody, painted by Paola Groeber, who painted only one in this color, c.1990. *Courtesy of Paola Groeber.*

Sorrel Peruvian Paso (resin, mold R2), Paola #810, "Elite El Paso," about 50-85 made, 1990; sculpture by Kathleen Moody. *Courtesy of Tina English-Wendt.* NM: $150.

Dark Bay Peruvian Paso (resin, mold R2), no #, Paola painted 13 in 1992 and sold them in December 1994; sculpture by Kathleen Moody. *Courtesy of Tina English-Wendt*. NM: $150.

Bay Miniature Horse (resin, mold R1), Paola #800, "Rich Bay—Heritage Hopeful," about 50 made, 1989; sculpture by Kathleen Moody. *Courtesy of Shirley Ketchuck*. NM: $85.

White Miniature Horse (resin, mold R1) with tawny mane and tail, Paola #801, "Heritage Mist," about 50 made, 1990; sculpture by Kathleen Moody. *Courtesy of Shirley Ketchuck*. NM: $85.

Pocket Pony Tennessee Walker (resin, mold R4) in Red Roan, Paola #860, "Trip Ticker," about 50 made, 1990; sculpture by Carol Gasper. *Courtesy of Carol Gasper*. NM: $75.

Pocket Pony Quarter Horse (resin, mold R3) in Dapple Grey, Paola #850, "Grey to Stay," about 50 made, 1990; sculpture by Carol Gasper. *Courtesy of Shirley Ketchuck*. NM: $75.

The 7" series Arabian Family (molds 28, 29 & 30) consists of a 6.75" high Stallion, 6.25" Mare, and 4.75" Foal.

Gray 7" Arabian Mare (with black points), styrene, 1962-1964. $9-15.

Gray 7" Arabian Stallion (with black points), styrene, 1962-1964. *Courtesy of Eleanor & Shay Goosens.* $10-17.

Gray 7" Arabian Foal (with black points), styrene, 1962-1964. $8-14.

Dapple Grey 7" Arabian Stallion, Tenite, Paola #511, 1988-1990. *Courtesy of Eleanor Harvey.* VG-NM: $22-35.

Dapple Grey 7" Arabian Mare, Tenite, Paola #511, 1988-1990. *Courtesy of Eleanor Harvey.* VG-NM: $20-30.

Dapple Grey 7" Arabian Foal, Tenite, Paola #511, 1988-1990. *Courtesy of Eleanor Harvey.* VG-NM: $17-24.

Blue-Gray 7" Arabian Mare (with white mane and tail), styrene, a 1960s color. *Courtesy of Sandy Tomezik.* $14-24.

Blue-Gray 7" Arabian Stallion (with white mane and tail), styrene, a 1960s color. *Courtesy of Peggy Howard.* $16-26.

Blue-Gray 7" Arabian Foal (with black mane and tail), styrene, a 1960s color. *Courtesy of Eleanor Harvey.* $12-20.

White-Grey 7" Arabian Stallion (with gray mane and tail, black lower legs, and eyes not painted), styrene, #6761, spring 1968. *Courtesy of Sandy Tomezik.* $8-14.

Buckskin 7" Arabian Stallion (with body color molded in and eyes unpainted), styrene, #6761, spring 1968. $8-14.

7" Arabian Family in white with black points, styrene, a variation on the usual gray-with-black points color, 1962-1964. Mare: $9-15; Stallion: $10-17; Foal: $8-14.

7" Arabian Family in Rose Grey, Tenite, Paola #510, 1987-1990. VG-NM—Mare: $20-30; Stallion: $22-35; Foal: $17-24.

Dark Rose Grey 7" Arabian Mare, Tenite, Paola, an early variation in very limited quantities, 1987. *Model, courtesy of Terry Davis*. VG-NM: $23-35.

7" Arabian Stallions in rose grey (*left*) and darker rose grey (very limited quantity, 1987), both by Paola, both Tenite. *Courtesy of Tina English-Wendt*. Darker Stallion: VG-NM: $25-38.

Dark Rose Grey 7" Arabian Foal, Tenite, very limited quantity, 1987. *Model, courtesy of Terry Davis*. VG-NM: $20-27.

Bay 7" Arabian Mare (with white stockings), styrene, a 1960s color. $12-18.

Bay 7" Arabian Stallion (with white stockings), styrene, a 1960s color. $14-20.

Bay 7" Arabian Foal (with white stockings), styrene, a 1960s color. $10-16.

Dark Sorrel 7" Arabian Mare, Tenite, Steven #271, "Liver Chestnut," 1993-1994. $13-23. *Courtesy of Eleanor Harvey.*

Dark Sorrel 7" Arabian Stallion, Tenite, Steven #271, "Liver Chestnut," 1993-1994. $15-28. *Courtesy of Eleanor Harvey.*

Dark Sorrel 7" Arabian Foal, Tenite, Steven #271, "Liver Chestnut," 1993-1994. $10-19. *Courtesy of Eleanor Harvey.*

Sorrel 7" Arabian Mare, Tenite, Paola, #213TC test color, 11 sets made, Christmas 1986. VG-NM: $23-35.

Sorrel 7" Arabian Stallion and Foal, Tenite, #213TC test color, 11 sets made, Christmas 1986. VG-NM—Stallion: $25-38; Foal: $20-27.

Root-Beer Bay 7" Arabian Stallion (with black lower legs, molded in golden-yellow plastic), a 1960s color that is not common. *Courtesy of Sandy Tomezik.* $17-30.

Root-Beer Bay 7" Arabian Mare and Foal (with black lower legs), a 1960s color; illustration is from a Hartland Plastics archive photo (*used with permission*). Mare: $15-25; Foal: $12-22.

Light Bay 7" Arabian Family, styrene, a Paola test color, Christmas 1986, about 21 sets. *Courtesy of Daphne R. Macpherson.* VG-NM: Mare: $22-33; Stallion: $24-35; Foal: $18-24.

Dark Grey 7" Arabian Family, Tenite, Paola #213TB, a Christmas 1986 test color, 19 sets made. VG-NM—Mare: $22-33; Stallion: $24-35; Foal: $18-24.

Golden Yellow (unpainted) 7" Arabian Mare, styrene, a budget model, 1966-1969. Mare: $8-13; matching Stallion (not shown): $10-18.

Golden Yellow (unpainted) 7" Arabian Foal, styrene, a budget item, 1966-1969. $8-13.

7" unpainted Arabian Mares in ivory-white (*left*) and pale taupe, styrene, budget models, 1966-1969. $8-13 each.

Pale Taupe (unpainted) 7" Arabian Stallion, styrene, a budget model, 1966-1969. $10-18.

Pale Taupe (unpainted) 7" Arabian Foal, styrene, a budget model, 1966-1969. $8-13.

The 7" series Arabian families include a Stallion, but the 5" series Arabian families do not. The mares and foals in the two sizes are similar in build and pose.

The 7" series Thoroughbred Family (molds 31, 32 & 33) consists of a 6.25" high Stallion, a 4.75" high grazing mare, and a crouching (nursing) foal that is 3.5" tall at the highest part of its tail.

Bay 7" Thoroughbred Stallion and Mare, styrene, #673, 1962-1964. Stallion: $9-15; Mare: $7-12.

Bay 7" Thoroughbred Foal, styrene, #673, 1962-1964. $6-10.

Very Dark Gray (unpainted) 7" Thoroughbred Stallion, styrene, a budget model, 1966-1969. $7-12.

Black (unpainted) 7" Thoroughbred Mare, styrene, a budget model, 1966-1969. $5-9.

Black (unpainted) 7" Thoroughbred Foal, styrene, a budget model, 1966-1969. *Courtesy of Sandy Tomezik.* $4-8.

Besides dark gray, this Hong Kong copy of the Hartland 7" series Thoroughbred Stallion is found in bay and palomino. *Model, courtesy of Jaci Bowman.* Copies: $5-9.

The Hong Kong copy (in palomino, *right*) is the same size as the Hartland 7" Thoroughbred Stallion (*left*). *Palomino copy, courtesy of Laura K. Whitney.* Copy: $5-9.

In the 7" and 5" series Thoroughbred Families, the mares and foals are similar in appearance, but only the 7" family has a stallion.

The 7" Appaloosa and Quarter Horse Families (each using molds 34, 35 & 36) consist of a 6.25" high Stallion, 5.5" Mare, and 4.5" Foal.

7" Appaloosa Stallion, styrene, #681, 1963-1964 and 1968. *Model, courtesy of Terry Davis.* $10-17.

7" Appaloosa Mare, styrene, #681, 1963-1964 and 1968. *Model, courtesy of Terry Davis.* $12-20.

7" Appaloosa Foal, styrene, #681, 1963-1964 and 1968. *Model, courtesy of Terry Davis.* $9-15.

7" Quarter Horse Stallion in buckskin with golden shadings, styrene, #672, 1962-1964. $9-15.

7" Quarter Horse Mare in buckskin with golden shadings, styrene, #672, 1962-1964. $10-17.

7" Quarter Horse Foal in buckskin with golden shadings, styrene, #672, 1962-1964. $8-14.

Buckskin 7" Quarter Horse Stallion (without golden shadings), #672, 1962-1964. $7-13.

Buckskin 7" Quarter Horse Mare (without golden shadings), #672, 1962-1964. $8-14.

Buckskin 7" Quarter Horse Foal (without golden shadings), #672, 1962-1964. $6-12.

7" series Quarter Horse Foals with golden body shadings (left) and without the shadings. Both are molded in golden yellow plastic, have dark shading on the face, and were made by Hartland Plastics in the 1960s.

Buckskin 7" Quarter Horse Stallion with no facial shadings, Durant #4400, 1970-1973. $4-7.

Buckskin 7" Quarter Horse Mare with no facial shadings, Durant #4400, 1970-1973. *Courtesy of Peggy Howard.* $5-8.

Buckskin 7" Quarter Horse Foal with no facial shadings, Durant #4400, 1970-1973. $3-6.

Two Bay 7" Quarter Horse Stallions from the #900 Silver Canyon Roundup Set, 1965-1966: model *at left* was painted bay over golden-yellow plastic; model *at right* was painted bay over reddish-brown plastic. *Courtesy of Sandy Tomezik.* Each: $12-25.

The Sorrel 7" Quarter Horse Mare from the Silver Canyon Roundup Set, 1965-1966, styrene plastic. *Courtesy of Eleanor Harvey.* $12-25.

Bay 7" Quarter Horse Foal from #900 Silver Canyon Roundup Set, 1965-1966. *Courtesy of Sandy Tomezik.* $10-20.

Chestnut 7" Quarter Horse Mare, Tenite, Paola #212TC, a Christmas 1986 test color, 18 sets with matching stallion and foal were made. *Courtesy of Sandy Tomezik.* VG-NM—Mare: $25-38; Stallion (not shown): $23-35.

Chestnut 7" Quarter Horse Foal, Tenite, Paola #212TC, a Christmas 1986 test color, 18 made; sold with matching stallion and mare. *Courtesy of Sandy Tomezik.* VG-NM: $20-27.

Golden-Yellow (unpainted) 7" Quarter Horse Stallion, styrene, a budget model, 1966-1969. *Model, courtesy of Sandra J. Truitt.* $5-10.

Golden-Yellow (unpainted) 7" Quarter Horse Mare, styrene, a budget model, 1966-1969. *Model, courtesy of Sandra J. Truitt.* $5-10.

Golden-Yellow (unpainted) 7" Quarter Horse Foal, styrene, a budget model, 1966-1969. $5-10.

Reddish Brown (unpainted) 7" Quarter Horse Stallion, styrene, a budget model, 1966-1969. $5-10.

Reddish Brown (unpainted) 7" Quarter Horse Mare and Foal, styrene, budget models, 1966-1969. *Courtesy of April and Jon Powell.* Each: $5-10.

A 5.25" high brass copy of the Hartland 7" Quarter Horse mare. On the copy, the right foreleg is raised; on the original, the left foreleg is raised. *Courtesy of Barri Mayse.* $6-12.

A 3.5" high brass copy of the Hartland 7" Quarter Horse Foal. On both the copy and the original, the left foreleg is raised. *Courtesy of Barri Mayse.* $5-10.

The 7" series Quarter Horse Mare (in buckskin, *right*) was copied in plastic in Hong Kong (in palomino, *left*). *Palomino copy, courtesy of Laura K. Whitney.* Copy: $4-6.

The palomino Hong Kong copy (*left*) of the Hartland 7" series Quarter Horse Foal is smaller than the Hartland model, and has thicker hind legs. *Palomino copy, courtesy of Laura K. Whitney.* Copy: $3-5.

In the Hong Kong palomino family, the stallion is 6.25" high, mare is 5.25" to the highest part of the neck, and the foal is just over 3". *Models, courtesy of Laura K. Whitney.* Entire set: $13-23.

The Hong Kong copies of the Hartland 7" Quarter Horse Mare and Foal were made in dark gray, too. *Courtesy of Judy & Kelley Harding.* Mare: $4-6; Foal: $2-4.

Like the Arabian and Thoroughbred families, the 7" series Quarter Horse Family has a 5" series counterpart that looks very similar but never had a stallion.

The 7" Morgan and Pinto Families (each using molds 37, 38 & 39): consist of a 6.5" high Stallion, 6.25" Mare, and 3.75" Foal.

Copper Sorrel 7" Morgan Stallion in Tenite plastic, from Hartland Plastics set #677, made during part of 1963-1968. $8-14.

Copper Sorrel 7" Morgan Mare in Tenite plastic, from Hartland Plastics set #677, made during part of 1963-1968. $8-14.

Copper Sorrel 7" Morgan Foal in Tenite plastic, from Hartland Plastics set #677, made during part of 1963-1968. $7-12.

Copper Sorrel 7" Morgan Stallion in styrene plastic, from Hartland Plastics set #677, made during most of 1963-1968. $6-10.

Copper Sorrel 7" Morgan Mare in styrene plastic, from Hartland Plastics set #677, made during most of 1963-1968. $6-10.

Copper Sorrel 7" Morgan Foal in styrene plastic, from Hartland Plastics set #677, made during most of 1963-1968. $5-9.

Red-Orange Sorrel 7" Morgan Stallion by Durant, styrene, from Durant set #4404, 1970-1973. $4-7.

Red-Orange Sorrel 7" Morgan Mare by Durant, styrene, from Durant set #4404, 1970-1973. $4-7.

Red-Orange Sorrel 7" Morgan Foal by Durant, styrene, from Durant set #4404, 1970-1973. $3-6.Photo 239-07

The Durant Morgan Mare (*right*) is red-orange with no shadings while the Hartland Morgan Mare (*left*) has abundant shadings.

7" Pinto Stallion in Tenite plastic, from Hartland Plastics set #680, made during part of 1964-1968. $8-14.

Left side of Tenite, 7" Pinto Stallion, #680.

7" Pinto Mare in Tenite plastic, from Hartland Plastics set #680, made during part of 1964-1968. *Model, courtesy of Sande Schneider.* $8-14.

Left side of another, Tenite, 7" Pinto Mare, #680.

7" Pinto Foal in Tenite plastic, from Hartland Plastics set #680, made during part of 1964-1968. *Model, courtesy of Sande Schneider.* $7-12.

Left side of Tenite, 7" Pinto Foal, #680. *Model, courtesy of Sande Schneider.*

7" Pinto Stallion in styrene plastic, Hartland Plastics #680, 1964-1968, and Durant #4404, 1970-1973. $6-10.

Left side of styrene, 7" Pinto Stallion, #680 and #4404.

7" Pinto Mare in styrene plastic, Hartland Plastics #680, 1964-1968, and Durant #4404, 1970-1973. $6-10.

Left side of styrene, 7" Pinto Mare, #680 and #4404.

7" Pinto Foal in styrene plastic, Hartland Plastics #680, 1964-1968, and Durant #4404, 1970-1973. $5-9.

Left side of styrene, 7" Pinto Foal, #680 and #4404.

Mahogany Bay (Pearled) 7" Morgan Family, styrene, Paola #211HP, a Christmas 1986 test color, 12 sets made. VG-NM—Mare: $25-35; Stallion: $25-35; Foal: $20-27.

Light Gray (unpainted) 7" Morgan Stallion, styrene, a budget model, 1966-1969. $5-10; matching Mare and Foal (*not shown*): $5-10 each.

Ivory-White (unpainted) 7" Morgan Stallion, styrene, a budget model, 1966-1969. *Courtesy of Carla S. Clifford.* $5-10.

Ivory-White (unpainted) 7" Morgan Mare, styrene, a budget model, 1966-1969. *Courtesy of Carla S. Clifford.* $5-10.

Ivory-White (unpainted) Morgan Foal, styrene, a budget model, 1966-1969. $5-10.

Reddish Brown (unpainted) Morgan Stallion, styrene, a budget model, 1966-1969. $5-10.

Reddish Brown (unpainted) Morgan Mare, styrene, a budget model, 1966-1969. $5-10.

Reddish Brown (unpainted) 7" Morgan Foal, styrene, a budget model, 1966-1969. $5-10.

The 7" Tennessee Walker Family (molds 40, 41 & 42)
includes a 6.5" high Stallion, 5.75" Mare, and 4.75" Foal.

Blue Roan 7" Tennessee Walker Stallion, Tenite, from Paola set #500, 1987-1990. VG-NM: $20-30.

Blue Roan 7" Tennessee Walker Mare, Tenite, from Paola set #500, 1987-1990. VG-NM: $20-30.

Blue Roan 7" Tennessee Walker Foal, Tenite, from Paola set #500, 1987-1990. VG-NM: $17-24.

White-Grey (Pearled) 7" Tennessee Walker Stallion, Tenite, from Paola set #500P, a 1989 special. *Courtesy of Judith Miller.* VG-NM: $20-30.

White-Grey (Pearled) 7" Tennessee Walker Mare, Tenite, from Paola set #500P, a 1989 special. *Courtesy of Sandy Tomezik.* VG-NM: $20-30.

White-Grey (Pearled) 7" Tennessee Walker Foal, Tenite, from Paola set #500P, a 1989 special. *Courtesy of Judith Miller.* VG-NM: $17-24.

Light Bay 7" Tennessee Walker Stallion, Tenite, from Paola set #501, 1988-1990. VG-NM: $20-30.

Light Bay 7" Tennessee Walker Mare, Tenite, from Paola set #501, 1988-1990. VG-NM: $20-30.

Light Bay 7" Tennessee Walker Foal, Tenite, from Paola set #501, 1988-1990. VG-NM: $17-24.

Dark Sorrel 7" Tennessee Walker Stallion, Tenite, from Steven set #271-20, "Liver Chestnut," a 1994 special for Black Horse Ranch. VG-NM: $20-30.

Dark Sorrel 7" Tennessee Walker Mare, Tenite, from Steven set #271-20, "Liver Chestnut," a 1994 special for Black Horse Ranch. VG-NM: $20-30.

Dark Sorrel 7" Tennessee Walker Foal, Tenite, from Steven set #271-20, "Liver Chestnut," a 1994 special for Black Horse Ranch. VG-NM: $17-24.

Palomino (Glossy) 7" Tennessee Walker Stallion, styrene, from Hartland Plastics set #684, 1965-1967. $12-25.

Bay Mare and Sorrel Foal, styrene, from the 7" Tennessee Walker Family set #684 by Hartland Plastics, 1965-1967. Mare: $12-25; Foal: $10-20.

The brown paint on some of the 1960s, 7" Tennessee Walker foals and mares turned greenish, a bit more olive green than this photo captures. *Courtesy of Susan Bensema Young, who calls hers, "Olivene" and "Jadite."* Mare: $8-16; Foal $7-14.

Palomino 7" Tennessee Walker Stallion, Tenite, Steven #269-20, a 1994 special for Black Horse Ranch. It has a matte or satin finish. VG-NM: $30-40.

Palomino 7" Tennessee Walker Mare and Foal, Tenite, Steven set #269-14, a special for the 1993 JCPenney Christmas catalog. Mare: $10-20; Foal: $8-15.

The 1960s (left) and 1990s (right) Palomino 7" Tennessee Walker Stallions are very different in color. Note the orange knees and hocks on the 1990s model.

Dark Bay 7" Tennessee Walker Stallion, Tenite, a Paola test color, #210TB, "Blood Bay," Christmas 1986; 24 stallion-mare-foal sets in this color were made. *Courtesy of Carla S. Clifford.* VG-NM: $23-35.

Dark Blue Roan (not pearled) 7" Tennessee Walker Stallion, styrene, Paola #210HB, a Christmas 1986 test color. About five stallion-mare-foal sets were non-pearled, dark blue roan; seven sets were non-pearled, light blue roan, also #210HB. VG-NM: $23-35 per adult; $20-25 per foal.

Light Blue Roan (Pearled) 7" Tennessee Walker Stallion, styrene, Paola #210HP, a Christmas 1986 test run of 19 stallion-mare-foal sets; there were no dark blue roan models with pearled finish. *Courtesy of Carla S. Clifford.* VG-NM: $23-35 per adult; $20-25 per foal.

Light Blue Roan (Pearled) 7" Tennessee Walker Mare, styrene, Paola #210HP, Christmas 1986. *Courtesy of Carla S. Clifford.* VG-NM: $23-35.

Light Blue Roan (Pearled) 7" Tennessee Walker Foal, styrene, Paola #210HP, Christmas 1986. *Courtesy of Carla S. Clifford.* VG-NM: $20-25.

Wedgewood (Blue) 7" Tennessee Walker Stallion, Tenite, Paola; a handful made, 1990. *Courtesy of Eleanor Harvey.*

Wedgewood (Blue) 7" Tennessee Walker Mare, Tenite, Paola; a handful made, 1990. *Courtesy of Eleanor Harvey.*

Wedgewood (Blue) 7" Tennessee Walker Foal, Tenite, Paola; a handful made, 1990. *Courtesy of Eleanor Harvey.*

The 7" series Saddlebred Families: The woodcut family (molds 43, 44 & 45) has a 6.5" high Stallion, 6.25" Mare, 4.75" Foal; the smooth (non-woodcut) family (molds 46 & 47) consists of a 6.5" high bay Mare and 5" sorrel Foal.

Walnut Woodcut 7" Saddlebred Stallion, Tenite, from Hartland Plastics set #685W, 1965-1967. $15-28.

Left side of Walnut 7" Saddlebred Stallion.

Walnut Woodcut 7" Saddlebred Mare, Tenite, from Hartland Plastics set #685W, 1965-1967. $15-28.

Left side of Walnut 7" Saddlebred Mare.

The walnut woodcut 7" Saddlebreds can be found in slightly lighter (*left*) and darker colors of tan plastic. The foal at right also shows more dark stain. $12-20 each.

Left side of Walnut Woodcut 7" Saddlebred Foal, Tenite, from Hartland Plastics set #685, 1965-1967.

Cherry Woodcut 7" Saddlebred Stallion, Tenite, from Hartland Plastics set #685C, 1965-1966. *Courtesy of Daphne R. Macpherson.* $20-35.

Cherry Woodcut 7" Saddlebred Mare, Tenite, from Hartland Plastics set #685C, 1965-1966. *Courtesy of Peggy Howard.* $20-35.

Cherry Woodcut 7" Saddlebred Foal, Tenite, from Hartland Plastics set #685C, 1965-1966. *Courtesy of Daphne R. Macpherson.* $15-25.

Ebony Woodcut 7" Saddlebred Stallion, Tenite, from Hartland Plastics set #685E, 1965-1966. *Courtesy of Carla S. Clifford.* $20-35.

Ebony Woodcut 7" Saddlebred Mare, Tenite, from Hartland Plastics set #685E, 1965-1966. This example has a shiny finish and no stain. *Model, courtesy of Jaci Bowman.* $20-35.

Ebony Woodcut 7" Saddlebred Foal, Tenite, from Hartland Plastics set #685E, 1965-1966. *Courtesy of Laura Diederich.* $15-25.

Bay 7" Saddlebred Mare with smooth surface (not woodcut), styrene, from Hartland Plastics set #764, 1965-1968. $12-25.

Sorrel 7" Saddlebred Foal with smooth surface (not woodcut), styrene, from Hartland Plastics set #764, 1965-1968. $9-15.

The woodcut, 7" Saddlebreds are a little smaller than the smooth, 7" Saddlebreds. *Top row:* woodcut mare and smooth mare. *Bottom row:* woodcut stallion, woodcut foal, and smooth foal. The smooth sets never had a stallion.

The 7" series Woodcut Thoroughbred Mare and Foal (molds 48 & 49) were a set. The Mare is 6.25" high, and the Foal is 4.5".

Ebony Woodcut 7" Thoroughbred Mare, Tenite, from Hartland Plastics set #683E, 1965 only. *Courtesy of Shirley Ketchuck.* $30-70.

Ebony Woodcut 7" series Thoroughbred Foal, Tenite, from Hartland Plastics set #683E, 1965 only. *Courtesy of Shirley Ketchuck.* $20-45.

Walnut Woodcut 7" Thoroughbred Mare, Tenite, from Hartland Plastics set #683W, 1965 only. $30-70.

Walnut Woodcut 7" series Thoroughbred Foal, Tenite, from Hartland Plastics set #683W, 1965 only. $20-45.

Cherry Woodcut 7" Thoroughbred Mare, Tenite, from Hartland Plastics set #683C, 1965 only. *Courtesy of Eleanor Harvey.* $30-70.

Cherry Woodcut 7" Thoroughbred Foal, Tenite, from Hartland Plastics set #683C, 1965 only. *Courtesy of Eleanor Harvey.* $20-45.

The 6" Arabian Stallion woodcut (mold 50) is 6" high.

Cherry Woodcut 6" Arabian Stallion with visible stain (factory antiquing), Tenite, Hartland Plastics #610C, 1965-1966. *Courtesy of Sandy Tomezik.* $10-25.

Shiny Cherry Woodcut 6" Arab Stallion molded in reddish-brown Tenite with no stain added, Hartland Plastics #610C, 1965-1966. $10-25.

Walnut Woodcut 6" Arabian Stallion (with stain), Tenite, Hartland Plastics #610W, 1965-1966. $10-25.

Walnut Woodcut 6" Arabian Stallion with shiny finish, molded in orange-tan plastic, Tenite, Hartland Plastics #610W, 1965-1966. $10-25.

Ebony Woodcut 6" Arabian Stallion (with matte finish), Tenite, Hartland Plastics #610E, 1965-1966. $10-25.

Shiny Ebony Woodcut 6" Arab stallion, Tenite, Hartland Plastics #610E, 1965-1966. $10-25.

The 5" Series Arabian Family (molds 51 & 52)
has two members: a 4.5" high Mare and 3.5" Foal.

Gray 5" Arabian Family, styrene, #6002, in 1963-1967 catalogs. *Courtesy of Eleanor & Shay Goosens*. Mare: $6-12; Foal: $4-8.

White 5" series Arabian Family with black points, styrene, Hartland Plastics #6002 in 1968 catalog; also #4201 in Durant 1970-1973 catalogs, but Durant sold leftover, 1960s stock before producing its own white 5" Arab family, with low black socks and no shading on the face. Mare: $5-10; Foal: $3-6; less with no shadings on face.

Bay 5" Arabian Family, styrene, series #6002, sold as a set, 1961-mid-1960s. Mare: $4-8; Foal: $3-6.

The Palomino 5" Arabian Family, styrene, series #6002, was included in the 1962 Red Bird catalog. Mare: $8-14; Foal: $6-12.

5" Arabian family with Bay Mare and Sorrel Foal, styrene, series #6002, sold as a set in the 1960s. Mare: $4-8; Foal: $6-10.

Buckskin 5" Arabian Family, styrene, series #6002, sold as a set in the 1960s. *Courtesy of Eleanor & Shay Goosens*. Mare: $6-12; Foal: $5-10.

The palomino 5" Arabian foal (*left*) was molded in brighter yellow plastic than the buckskin foal (*right*). *Palomino, courtesy of Cecile Bellmer.*

The Golden Yellow (unpainted) 5" Arabian Mares and Foals came in two shades of golden yellow plastic; they were styrene and sold as budget items, 1966-1969. Mares: $3-6 each; Foals: $2-4 each.

Reddish Brown (unpainted) 5" Arabian Mare and Foal, styrene, sold separately as budget models, 1966-1969. Mare: $4-6; Foal: $3-5.

Pale Taupe (unpainted) 5" Arabian Mare and Foal, styrene, sold separately as budget models, 1966-1969. Mare: $4-6; Foal: $3-5.

The 7" and 5" Arabian Mares were each made in the ivory-white and pale taupe unpainted colors. They were budget items sold individually from counter-top, cardboard bins in dime stores in 1966-1969.

Compare the Hartland 5" series Arabian Mare in buckskin (*left*) to the 5" toy horse (*right*) marked, "Nylint Corp/Made in/Hong Kong." Both are styrene. Nylint: $3-5.

A 2.5" high, white toy horse marked, "Made in Hong Kong," is a faithful copy of the Hartland 5" series Arabian Mare, shown here in golden yellow. Copy: 25 cents-50 cents.

The Hartland 5" series Arabian Foal in buckskin (left) has a 3.5" Nylint counterpart (right) marked, "Nylint Corp/Made in/Hong Kong." Both are styrene. Nylint: $2-4.

The unmarked copies of the Hartland 5" series Arabian foal have been found in electric yellow, white, and black. They are made of styrene, just like the Hartland, and were probably made in Hong Kong. Copies: $1-2 each.

Hartland made Gray Arabian mares in both the 7" and 5" series. Both have points painted black and bodies shaded with gray paint over white plastic.

These copies or take-offs on the Hartland 5" series Arabian mare measure 1.75" to 2.5" high. The toy at left is marked, "Made in Hong Kong," but the other three—white, light gray, and red-brown—are not marked. All are made of a rubber-like plastic. Each: 25 cents-50 cents.

A close copy of the Hartland 5" series Arabian foal has eyes that slant upward and legs that are thinner above the fetlocks. The Hartland in golden yellow (left) is marked "Hartland Plastics Inc." The copy in electric yellow (right) is not marked in any way. Copy: $1-2 each.

This unmarked copy of the Hartland 5" Arab Foal has thicker forelegs and a head that reminds me of a chess knight. It is styrene and just under 3.5" high. Copy: $1-2 each.

Likewise, the Gray Arabian Foals appear in both the 7" and 5" series. The 7" Gray Arabians have a protective gloss coat, but the 5" Gray Arabians do not and their gray shadings can rub off. Wash them gently or not at all.

The 5" series Thoroughbred Family (molds 53 & 54) has two members: a 3.75" high grazing Mare and 2.75" high crouching (nursing) Foal measured at the top of its tail.

Bay (Glossy, Shaded Maroon Bay) 5" Thoroughbred Family, styrene, in Hartland Plastics series #6002, 1963-1969, and some (leftover, 1960s stock) were sold by Durant (as series #4201) in the earlier part of 1970-1973. Mare: $4-6; Foal: $2-4.

Unshaded Bay 5" Thoroughbred Family, styrene, not glossy, with low black socks; produced by Durant (as #4201) during 1970-1973 after it sold out the leftover, shaded bays. *Foal, courtesy of Cecile Bellmer.* Mare: $4-6; Foal: $2-4.

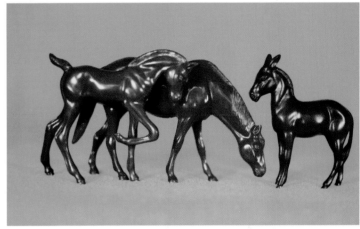

Cranberry Red (unpainted) 5" Thoroughbred Mare and Foal, styrene, sold separately as budget items, 1966-1969. Their color, dark red with a hint of purple-red (magenta), is difficult to capture on film. Mare: $4-7; Foal: $3-5.

Three, related colors of unpainted, 1960s budget models, *from left*: reddish brown (7" Morgan Foal), cranberry red (5" Thoroughbred Mare), and plum brown (Farm Donkey). Here, the cranberry red model looks more like magenta and less like dark red than it usually does to the eye.

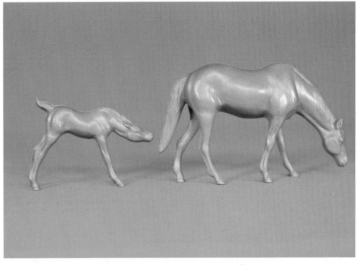

Golden Yellow (unpainted) 5" series Thoroughbred Mare, styrene, a budget item, 1966-1969. $3-5.

Golden Yellow (unpainted) 5" Thoroughbred Foal and Mare, styrene, sold separately as budget items, 1966-1969. Mare: $3-5; Foal: $2-3.

The Bay 7" and 5" series Thoroughbred Mares are nearly identical except for their heights: 4.75" versus 3.75". They are measured at the highest point, the withers, which is the bony ridge where the neck joins the back.

The 5" and 7" series Bay Thoroughbred Foals strike the same nursing pose, but the 5" series foal's tail points more directly up, and the 7" series foal's head is tilted up more, with his nose so high as his ears. The heights, measured at the highest part of the tail, are 2.75" and 3.5".

The 5" series Quarter Horse family (molds 55 & 56) has a 4.25" high Mare and 3" Foal.

Buckskin 5" Quarter Horse Mare with shadings, styrene, from Hartland Plastics series #6002, 1963-1969; also in Durant 1970-1973 catalogs (as #4201), but the ones with shadings were leftover, 1960s stock. $6-10.

Buckskin 5" Quarter Horse Foals with lesser (left) and greater amounts of caramel shadings; both from 1963-1969. Foals produced in the later part of 1970-1973 would have low socks and no face or body shadings. With shadings: $4-8; without shadings: $2-4.

Buckskin 5" Quarter Horse Mare without shadings and with low socks, styrene, Durant series #4201; this is the model Durant produced during 1970-1973 after the shaded buckskins sold out. $4-7.

Bay 5" Quarter Horse Mare, styrene, from Hartland's Sunny Acres Farm series, sold on a card with a matching foal, 1965. The model is molded in brownish-red plastic with black points and body shadings. *Courtesy of Sandy Tomezik.* $6-10.

The same Bay 5" Quarter Horse Mare can, under some lighting/viewing conditions, appear a luminous, "root beer" color. The 1960s, 9" maroon bay Thoroughbred, #873, can look like this, also. $6-10.

The Bay 5" Quarter Horse Foal, styrene, went with the bay mare in the Sunny Acres Farm series, 1965. $4-7.

Buckskin Quarter Horse Mares are found in both the 7" and 5" family series. Their heights, 5.5" and 4.25", are taken at the highest part of their arched necks. The tail is more blunt on the 7" mare; on the 5" mare, the right forefoot is more turned up so that the bottom of the hoof is almost parallel to the ground.

The 7" buckskin Quarter Horse Foal holds his head higher than the 5" Quarter Horse Foal. The 5" Quarter Horse Foal's pose is similar to that of the 7" series Morgan-Pinto Foal.

Hartland's Farm Animals included a 4.25" high Horse (mold 57) and a 3.75" Donkey (mold 58); both were always unpainted. A 4.5" Donkey (mold 59) belonged to the Nativity Set.

Pale Taupe (unpainted) Farm Horse, styrene, a budget item, 1966-1968. $5-10.

Farm Set Horses in five colors of unpainted styrene: cream, white, golden yellow, pale taupe, and plum brown; from 1966-1968, and some could be from 1970-1973 and/or 1985-1986. Each: $5-10.

Farm Horses in white and pale taupe styrene. The white one, with visible glue on the tail, is probably by Durant, 1970-1973. Each: $5-10, but $3-6 if visible glue.

I think the Farm Horse, in golden yellow, is a Saddlebred. Compare it to the black, Saddlebred-like horse dating to 1961 Nylint catalogs. Hartland's Arabs have chins and tails lifted higher. *Nylint horse, courtesy of Cecile Bellmer.* Nylint: $3-6.

The Farm Donkey in plum brown, unpainted styrene, a budget item, 1966-1968. $4-6.

Farm Donkeys in pale gray and golden yellow (unpainted styrene), budget items, 1966-1968. $4-6.

White, Pale Taupe, and Cream Farm Donkeys, 1966-1968; the white donkey with yellow glue visible is probably from 1970-1973. Some Farm Donkeys were also made in 1985-1986. $4-6.

The Nativity Set Donkey stands with one hind leg relaxed. It's Tenite, 4.5" high, usually painted dark gray or in unpainted white, and was made for a few years in the early 1950s. $8-15.

Chapter 6: **Tinymite Horses**

The Tinymite Horses came in six breeds (molds 60-65) and, with mold variations, came in 13 different shapes, all between 2.25" and 2.75" high.

Tinymite Arabians in sorrel: old mold with separated ears and curled tail, 1965 *(left)*, and new mold with joined ears, 1966-1969; both styrene. Old mold: $10-16; new mold: $8-12.

Left side of Tinymite Arabians in sorrel, in the old *(left)* and new mold versions; both styrene.

Tinymite Arabian in sorrel, an intermediate mold with separated ears and straight tail, 1965; styrene. *Courtesy of Eleanor Harvey.* $10-16.

Tinymite Arabian in orange sorrel by Durant, 1970-1973; from the new mold with joined ears; styrene. $6-10.

Tinymite Belgians in bay: old mold with separated ears, 1965, *(left)* and new mold with joined ears, 1966-1969; both styrene. *Models, courtesy of Terry Davis.* Old mold: $12-20; new mold: $10-16.

Tinymite Belgian in orange by Durant, 1970-1973; from the new mold with joined ears; styrene. *Courtesy of Eleanor Harvey.* $4-8.

Tinymite Thoroughbred in buckskin; the old mold with separated ears, 1965; styrene. *Courtesy of Eleanor Harvey.* $10-20.

Both of these black, Tinymite Thoroughbreds are the new mold with joined ears: by Hartland Plastics, 1966-1969 with letter "I" on inside leg *(left)* and by Durant with letters "LJ" on inside leg, 1970-1973. Both are styrene. Each: $6-12.

Tinymite Quarter Horse in black; the old mold with separated ears, 1965; styrene. *Courtesy of Eleanor Harvey.* $10-20. Not shown: Black Quarter Horse, new mold (with joined ears), by Durant, 1970-1973. $6-12.

Both of these buckskin, Tinymite Quarter Horses are the new mold with joined ears: by Hartland Plastics, 1966-1969 *(left)*, and by Durant, with low socks, 1970-1973 (right). Both are styrene. *Hartland model, courtesy of Terry Davis.* 1960s: $8-16; 1970s: $7-14.

Tinymite Tennessee Walkers: old mold with separated ears, 1965 *(left)*, and new mold with joined ears, 1966-1969; both are by Hartland Plastics, in white with black points. *Old mold, courtesy of Terry Davis.* Old mold: $10-14; new mold: $8-12.

Tinymite Tennessee Walker in white with black points by Durant, 1970-1973; styrene; new mold, with joined ears. $7-10.

Tinymite Morgan in palomino, old mold, 1965; styrene; note wavy bottom edge of mane. *Courtesy of Shirley Ketchuck.* $10-16.

Tinymite Morgans in palomino, both the new mold: by Hartland Plastics, 1966-1969 *(left)* and by Durant Plastics, 1970-1973; both are styrene; note smooth, bottom edge of mane. The Durant palomino has paler color. 1960s: $8-14; 1970s: $6-10.

Tinymite Morgan (new mold) by Durant in black (factory-painted black over white plastic). *Courtesy of Eleanor Harvey.* $8-16.

Tinymite Arabian, old mold in unpainted, brown Tenite, 1960s. *Courtesy of Daphne R. Macpherson.* $6-10.

This Arabian in unpainted black is similar to the old mold, but has no assembly letters, and is evidently a copy. *Model, courtesy of Terry Davis.* $10-18.

Tinymite Arabian, new mold in unpainted, brown Tenite, 1960s. $4-8.

Tinymite Belgian, old mold in unpainted, golden-yellow Tenite; 1960s. *Courtesy of Eleanor Harvey.* $6-10.

Tinymite Belgian, new mold in unpainted, white styrene, 1960s. *Model, courtesy of Jacqueline Tierney.* $8-14.

Tinymite Belgian, new mold in unpainted brown Tenite, 1960s. *Model, courtesy of Jackie Himes; photo, courtesy of April and Jon Powell.* $4-8.

Tinymite Thoroughbred, old mold in unpainted, golden-yellow Tenite, 1960s. *Courtesy of Daphne R. Macpherson.* $6-10.

Tinymite Thoroughbred, new mold in unpainted brown Tenite, 1960s. $4-8.

Tinymite Thoroughbred *(left)* and Quarter Horse, both new molds, in unpainted golden-yellow Tenite. *Models, courtesy of Terry Davis.* Each: $4-8.

Tinymite Quarter Horse, new mold, in unpainted, white styrene, 1960s. *Model, courtesy of Jacqueline Tierney.* $8-14.

Tinymite Quarter Horse, old mold, in unpainted, brown Tenite, 1960s. *Courtesy of Daphne R. Macpherson.* $6-10.

Tinymite Tennessee Walker, old mold, in unpainted, brown Tenite, 1960s. *Model, courtesy of Jackie Himes; photo, courtesy of April and Jon Powell.* $6-10.

Tinymite Tennessee Walker, old mold, in unpainted, golden-yellow Tenite, 1960s. *Courtesy of Daphne R. Macpherson.* $6-10.

Tinymite Tennessee Walker *(left)* and Morgan, both in unpainted brown Tenite. The Walker is the new mold; the Morgan is the old mold; both 1960s. New mold: $4-8; old mold: $6-10.

Right side of Tinymite Morgan, old mold, in unpainted, golden-yellow Tenite, 1960s. *Courtesy of Daphne R. Macpherson.* $6-10.

Tinymite Morgans in unpainted, golden yellow Tenite: old mold *(left)* and new mold, 1960s. Old mold: $6-10; new mold: $4-8.

A copy of the Tinymite Thoroughbred, in brown with black mane and tail. *Courtesy of Daphne R. Macpherson.* $3-6.

A copy of the Tinymite Belgian, in brown with black mane and tail. Note the different profile on the tail. *Courtesy of Daphne R. Macpherson.* $3-6.

The Hartland Tinymite Belgian, old mold in bay, has body shadings, black lower legs, finer ears, and a curved tail profile. *Courtesy of Sandy Tomezik.* $12-20.

Compare the orange, Durant-Hartland Tinymite Belgian with a copy of the new-mold Belgian painted dark gray-brown. The copy has different texture lines on the mane, and no letters on the inside leg. Both are styrene. Durant: $4-8; copy: $3-6.

These copies of the Tinymite Morgan are marked with H. K. (for Hong Kong) and a "P" in a diamond. Their light brown and dark brown body colors are molded in. *Models, courtesy of Cecile Bellmer.* Each: $3-6.

These copies of the Tinymite Morgan and Arabian have no mark of any kind; they date to about 1988, and were probably made in Hong Kong. Each: $3-6.

Metal copies of the Tinymite Morgan exist in two sizes: 1.75" high and 2.5" high. *Courtesy of Barri Mayse.* $3-6 each.

Chapter 7: **Hartland Dogs**

There are 16 shapes of Hartland dogs, molds D1-D16.

Bullet, Roy Roger's German Shepherd, was sold for 50 cents from 1956-1963. He's Tenite, 4.25" high, with his name misspelled on the neck tag. Bullet is mold-marked with the Hartland name. *Model, courtesy of Jaci Bowman.* $20-35.

The Lying Down, German Shepherd, about 6" long and with a hollow bottom, was made in the 1950s by Hartland Plastics. It was intended as the first in a series of dog breeds, but did not sell well. *Model, courtesy of Jaci Bowman.* $25-50.

The Lying Down, German Shepherd is scaled to a larger size than Bullet. This example of the Lying Down dog is less fully painted. *Models, courtesy of Sande Schneider.* $20-45.

Tinymite Barkies, *top row*: Std. Poodle, Collie, and German Shepherd; *bottom row*: English Pointer, German S. H. Pointer, and St. Bernard; all from about 2"-2.5" high, Hartland Plastics, 1966-1969, and Durant, 1970-1973. (Durant colors include a black St. Bernard, mostly orange St. Bernard, and a German Pointer with orange spots.) Each: $8-15; $5-10 for Durant.

Tinymite Barkies, *top row*: Cocker Spaniel, Golden Retriever, and Beagle; *bottom row*: Irish Setter, Chesapeake Retriever, and Black Labrador (Retriever); all from about 2"-2.5" high; Hartland Plastics, 1966-1969, and Durant, 1970-1973. (Durant colors include an orange Irish Setter.) Each: $8-15; $5-10 for Durant.

The Tinymite Collie *(left)* is 2.25" high; the copy marked, "Hong Kong," is one-quarter inch taller. The copy is shown in the 1968 Nylint catalog with Nylint's #1710 Pet Mobile truck set. Copy: $4-8.

The Tinymite English Pointer *(left)* has a taller counterpart marked "Hong Kong," that came with the Nylint Pet Mobile Truck set, 1968. The third and final dog copied by Nylint is the Beagle. Copy: $4-8.

The Farm Dog and Puppy were part of Hartland's Sunny Acres Farms series, 1964-1965. The dog is 2.25" high, and the puppy is just under 2". They are styrene and in the same scale as the Tinymites. Dog: $8-15; puppy: $6-12.

Copies of the Hartland Farm dogs are the same size and marked, "Nylint Hong Kong." They have painted eyes and mouth, and a different pattern of brown over their head, back, and tail. Dog copy: $4-8; puppy copy: $3-6.

The Tinymite Barkies, including the Beagle *(left)* and Poodle, were #6007 when sold in individual boxes, and #6009 boxed in sets of three by Hartland Plastics, 1966-1969. *Models, courtesy of Jaci Bowman.* Box without cellophane: $1-2; with cell. intact: $2-3; then add the value of the model(s).

COLLECT 'EM ALL

ST. BERNARD • ENGLISH POINTER • BEAGLE
GERMAN SHORT HAIR POINTER • COLLIE
CHESAPEAKE RETRIEVER • COCKER
GOLDEN RETRIEVER • STANDARD POODLE
LABRADOR • GERMAN SHEPHERD • SETTER
LAP AND SPORTING DOGS

No. 6007 — No. 6009 in sets of 3

Authentic, Exact Models by

Hartland HARTLAND PLASTICS, INC.
HARTLAND, WISCONSIN

The back of the Tinymite Barkies box lists the 12 breeds of "lap and sporting dogs." *Box, courtesy of Jaci Bowman.*

Among the 1960s horses in the 9" and 11" series, some had eye whites painted in and some did not. The 9" Thoroughbred in Budweiser Brown had eye whites. *Courtesy of Sandy Tomezik.*

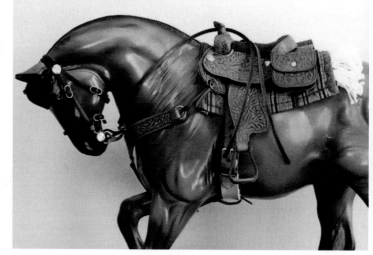

Among the 1980s-1990s horses (by Steven and/or Paola), some had eye whites and some did not. Paola's palomino Polo Pony had eye whites and a narrow blaze. *Courtesy of Eleanor & Shay Goosens.*

From 1987-1989, Paola sold hand-tooled, leather saddles by Kathleen Bond. Saddles and accessories, were available for 11" series (illustrated) and 7-9" series models. *Courtesy of Eleanor Harvey.* Original price: $50 per saddle, plus accessories; now worth about twice that.

The sticker on some 1960s woodcut models reads, "Wood Carved/Hand Rubbed/Finish/Molded of Tenite," but the originals were clay, not wood. *Model, courtesy of Sande Schneider.*

Paola's Pearl White 11" Quarter Horses are signed, "1991/HC/Hartland Collectables/Paola Groeber," and were numbered between 1 and 12. *Courtesy of Daphne R. Macpherson.*

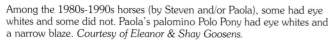

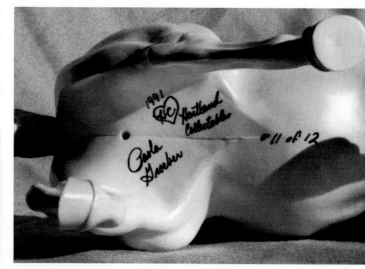

This #242 Dapple Grey, styrene, Steven Mustang shows crackling in the gray paint on its hips and legs; it also has more white on the neck than many. *Model, courtesy of Donna Anderson; photo, courtesy of Ellen W. Vogel.*

Where many of the Bright Sorrel, Steven, Jewel & Jade sets were banded inside their box, the reddish paint "melted" off to reveal the golden-yellow layer of paint below. Note the neck area.

The heat damage is more obvious on the left side of the Bright Sorrel Jewel & Jade set. About 15% of the 1993-1994 Steven models had defective paint that could turn sticky at about 82 degrees.

This Shaded Buckskin: Sandy Bay 9" Thoroughbred by Steven, 1988-1990, turned greenish, but reports of color changes in 1980s models are rare. *Courtesy of Sandy Tomezik.*

Do not use "coffee hot" water on models molded in colored plastic to try to straighten their legs. It permanently lightens the plastic. Note the right foreleg on the closer 7" series, woodcut Saddlebred Mare.

Orange-tan was the base paint for the 1960s, 9" Thoroughbreds in the Budweiser Brown color. The unpainted area on the tail is where the model was held by a clamp during painting.

121

This 9" Five-Gaiter went home with a 1960s, Hartland employee after the red paint coat, but before black points and black shading would have made it a #8679 Red Bay. It also looks more tomato red than the usual apple red color. *Model, courtesy of Jaci Bowman.*

The left half of this 9" Arab shows the copper paint used for the 1960s, #8679 Copper Sorrel model. His right half is bare, orange-tan plastic, but many were molded in white. This model and other unfinished/unusual models turned up at Hartland-area garage and estate sales in the past 20 years. *Model, courtesy of Jaci Bowman.*

This 7" Appaloosa Stallion escaped from the factory in the 1960s after getting his spots and black points, but before the gray body color and gloss coat were added. *Model, courtesy of Jaci Bowman.*

Likewise, this 7" series Appaloosa Mare from the 1960s turned up in the Hartland area. *Model, courtesy of Jaci Bowman.*

Unfinished, Hartland models also tend to turn up in Oklahoma, where the Durant factory made them in the 1970s. This caramel 11" Quarter Horse would have been a buckskin. *Courtesy of Charlene Marshall.*

This ceramic copy of the 11" series Quarter Horse was poured by collector Charlene Marshall in 1983 using a pour-it-yourself mold from Brush Country Molds, Texas. She cast, painted, and sold less than 10, but those poured by others increase the total. It's about 7.5" high. *Courtesy of Charlene Marshall.*

In 1983, Charlene Marshall cast 100 horses from the Brush Country mold and added a different mane and tail to each. Calling them Andalusians, she sold 50 in shaded white, 50 in bay, and one red chestnut for $50. They are signed and numbered. *Courtesy of Charlene Marshall.*

In the 1980s, Steven made a few test shots of the Farm horse in Tenite. They are brown, white, or black with non-smoothed seams. The regular, 1980s farm horses were styrene. *Courtesy of Sandy Tomezik.*

Steven also made a few, 1980s test shots of the Morgan *(illustrated)* and Arabian 7" series families molded in black Tenite. *Courtesy of Sandy Tomezik.*

In the 1960s, the large (9-11" series) horses were not, except for some woodcuts, sold unpainted, but a few survive that were rejects or pilfered. This 9" Saddlebred in white Tenite was found near Hartland, Wis. *Model, courtesy of Jaci Bowman.* $15-25+.

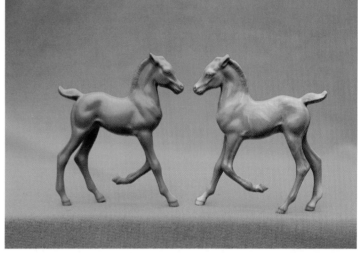

In the 1960s, 7" series and smaller horses were commonly sold unpainted, molded in various colors of styrene plastic. The golden-yellow 7" series Quarter Horse Foal *(left)* is common, but the one at right is molded in swirled-color plastic: mostly golden-yellow and white. Regular: $5-12; Swirled: $7-18.

The #883 Dark Bay, 1960s Polo Ponies were often molded in red-brown Tenite, as seen in this unpainted, unfinished model found near Hartland. Employees were allowed to take home unpainted rejects, which often had a bent leg, gouge, or chipped ear. *Model, courtesy of Jaci Bowman.* Flawed models: $8-15; $15-25+ if no obvious flaws.

This 7" series Saddlebred Mare and Foal molded in (white) Tenite are unusual because the Bay Mare/Sorrel Foal sets are "always" styrene. They were not sold as unpainted models. This unfinished pair escaped from the factory in the 1960s. *Model, courtesy of Jaci Bowman.*

In the 1980s, on the other hand, unpainted white models in the 9-11" sizes could be ordered from Paola's catalog. This is #239U, the 9" series Arab Mare in styrene, 1986. *Courtesy of Eleanor & Shay Goosens.* $15-25.

Paola advertised the unpainted, white 9-11" models as possibly having rough or mismatched seams that would need sanding. This is #BOD11, the 9" Thoroughbred in Tenite, 1988-1990. *Courtesy of Eleanor & Shay Goosens.* $20-$35.

The 11" Quarter Horse in unpainted, white Tenite was Paola #BOD3, 1988-1990. *Model, courtesy of Donna Anderson; photo, courtesy of Ellen W. Vogel.* $25-40.

Lady Jewel and Jade in unpainted, white Tenite were never officially for sale by Paola or Steven, but a few collectors managed to buy them. *Courtesy of Tina English-Wendt.* $50-80.

Collector and artist Susan Bensema Young restored the copper sorrel color on this 1960s, 11" series Five-Gaiter. *Courtesy of Susan Bensema Young.*

The owner said she has had this unusual, bronze-colored 11" series Arabian with white points "forever." *Courtesy of Laura Diederich.*

This is the only known example of a 7" Arabian family painted buckskin over white plastic. The mare's black stockings *(left)* don't match the other two models. *Courtesy of Laura Diederich.*

This Polo Pony painted silver came from a 1960s, Hartland employee. A 1960s Hartland official said there would have been no factory purpose for a model to be painted silver. *Model, courtesy of Jaci Bowman.*

"Driftwood" Mustang: This unusual model found in 1999 in an antique store in southeastern Wisconsin looks like factory work. It was molded in dark brown plastic, painted cream, then expertly airbrushed with dark shading.

This 9" Five-Gaiter from the 1960s is painted pearl white over the normal, "white with gray mane/tail" color in Tenite. It was found in the Hartland area and was probably a factory experiment. *Courtesy of Jaci Bowman.*

Unusual 5" series Thoroughbred Foal from the 1960s, found near Hartland: It's painted pearl brown over white plastic, and is probably unique. *Courtesy of Jaci Bowman.*

Boxes are presented here in order of model size, from the 11" series to Tinymites. Within each size group of models, the 1960s boxes appear first, followed by 1970s, 1980s, and 1990s boxes as applicable.

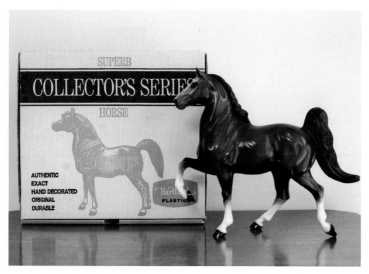

This 1967 Cherry Red Bay 11" Saddlebred came in a "Superb/Collector's Series/Horse" box picturing the 11" series Arabian. *Courtesy of Eleanor Harvey.* 1960s, 11" series box, complete with opaque, front panel: $8-20.

The 1968, Sorrel 11" Arabian in its cellophane-front box. *Model, courtesy of Jaci Bowman.* 1960s, 11" series box without cellophane: $6-12; if box is still sealed, add $5, then add the value of the model.

The side panel of the "Collector's Series" box for the Sorrel 11" Arabian. *Box, courtesy of Jaci Bowman.*

The box is stamped with the number of the 1967 Pearl White 11" Arabian: #99151. *Box, courtesy of Jaci Bowman.*

A 1967, 11" Arabian in Pearl White, still sealed in the box, is truly, as the box says, "Superb." *Model, courtesy of Jaci Bowman.* 1960s, 11" series box without cellophane: $6-12; if box is still sealed: add $5, then add the value of the model.

In 1983-1986, the 11" series models (by Steven) had a collapsible, cardboard box with the model's picture on the top half. This box reads, "Bay Quarter Horse No. 200." Box: $2-5.

This 1983-1986 Steven box held a "Saddlebred Bay Horse No. 202." Box: $2-5.

The sides of the 1983-1986 boxes for the 11" series illustrated one each of the Saddlebred, Arabian, and Quarter Horse molds.

The 1960s Copper Sorrel Five-Gaiter had this dark green box with cellophane front and blue-sky backdrop; the model's price mark is $2.00. *Courtesy of Judith Miller.* 1960s, 9" series box without cell: $5-12; box with intact cell: $8-15.

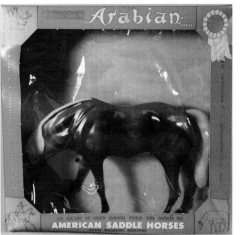

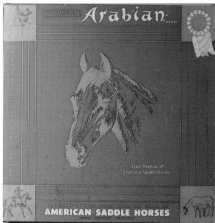

The back side of the 1960s, 9" series boxes read, "Exact Replicas of Champion Saddle Horses!/An Exact 10" model from the series of American Saddle Horses/Roger Williams, Sculptor." This one held the 9" Saddlebred (Five-Gaiter). *Box, courtesy of Jaci Bowman.*

The 1960s, blue-gray Arabian Mare had a blue-green box. The line drawings on three corners of the box were different for each breed. *Boxed model, courtesy of Jaci Bowman.* Box without cellophane: $5-12; box with intact cell: $8-15; model still fastened inside: add $5, then add the value of the model.

The back of the 1960s, blue-gray Arabian's box is the same as for the Saddlebred except for the breed name at the top and the line drawings in gold at the three corners. *Box, courtesy of Jaci Bowman.*

The #873, 9" Bay Thoroughbred, 1963-1967, has a burgundy box with sketches depicting polo, racing, and jumping at the corners. 1960s, 9" series box without cellophane: $5-12; box with intact cell: $8-15; model still fastened inside: add $5, then add the value of the model.

This 1960s box holds an ebony Tennessee Walker, 1964-1965. Gold fleurs-de-lis decorate the outside; the price mark is $1.88. *Boxed Model, courtesy of Sande Schneider.* Box without cellophane: $5-12; box with intact cell: $8-15; model still fastened inside: add $5, then add the value of the model.

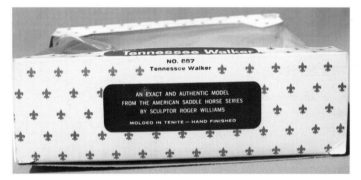

The bottom of the Tennessee Walkers' box reads, "No. 887/Tennessee Walker/An Exact and Authentic Model from the American Saddle Horse Series/by Sculptor Roger Williams/Molded in Tenite—Hand Finished." *Box, courtesy of Sande Schneider.*

The 1960s, "Woodgrain 3-Gaited Saddlebred" had a blue-and-white box decorated with a pattern of images of the six, non-woodcut horses in the 9" series. *Courtesy of Judith Miller.* Box without cellophane: $6-12; box with cell. intact: add $5, then add the value of the model.

The Red Bay, 9" Five-Gaiter, 1967-1968, appears in the blue-and-white, horse pattern box. *Boxed model, courtesy of Jaci Bowman.* Box without cellophane: $6-12; box with cell. intact: add $5, then add the value of the model.

A small panel of the Red Bay, Five-Gaiter's box reads, "Hartland Collector's Series/Authentic Hand Painted Models/Molded in Tenite." *Box, courtesy of Jaci Bowman.*

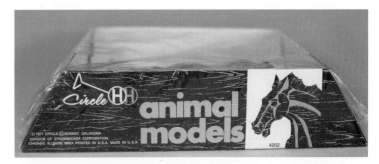

The bottom panel of the Durant Three-Gaiter's box reads, "Circle H Animal Models, #4202, copyright 1971 Circle H Durant, Oklahoma/ Division of Strombecker Corporation/Chicago, Illinois 60624 Printed in USA Made in USA."

This box dated 1971 holds the 9" series, woodcut Three-Gaiter. The box lists the Mustang and Tennessee Walker, but only the Three-Gaiter and woodcut Mustang are in the 1970-1973 Durant catalogs. Box without cellophane: $6-12; box with cell. intact: add $5, then add the value of the model.

Models in 1984-1989 picture frame boxes came with a paper insert identifying the model, in this case, the #239 Arabian Mare (in Dove Gray). Box: $2-5, then add the value of the model.

The 9" series horses made by Steven Mfg. from 1984-1989 usually appear in a picture-frame box, so that the model is visible from both sides. This is the #229 Polo Pony in blue roan. Box: $2-5, then add the value of the model.

In 1990, Steven Mfg. switched to a clam shell package with no cardboard on the outside, but cardboard inserts. This is a #249 Light Dapple Grey Arab Mare. Box: $2-5, then add the value of the model.

In the clam shell package, a "collector series card" names and describes the model: the #249 Arab Mare is "Silver Vanity." Another cardboard insert reads, in part, "Hartland Champions, #268-600." The total value of the cardboard inserts: $1-3.

129

Left: Steven boxed the Three-Stallion set for the 1991 JCPenney Christmas catalog in a plain box reading, "1 No. 278 Horse Assort., Penny's Cat. No. 671-5205, from Steven Mfg. Co., Hermann, MO," etc. (Penney's was spelled wrong.) Box: 50 cents-$2.

Right: The Steven, Three-Mare set for the 1992 JCPenney catalog traveled in this box identifying the set as Steven No. 279-05 and No. 671-0719 in the Penney's catalog. *Courtesy of James W. and Sandra J. Truitt.* Box: 50 cents-$2.

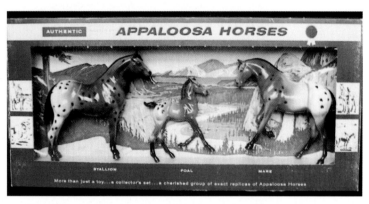

The 1960s, 7" series Appaloosa family box reads, "More than just a toy...a collector's set...a cherished group of exact replicas of Appaloosa Horses." *Courtesy of Heather M. McCurdy.* Box without cellophane: $5-12; with cell. intact: $8-15; models still fastened in box: add $5 plus the value of the models.

1960s 7" series Quarter Horse family box with the cellophane-front cover on. *Box, courtesy of Terry Davis.* Box without cellophane: $5-12; with cell. intact: $8-15; models still fastened in box: add $5 plus the value of the models.

The 1960s, 7" series Tennessee Walking Horse family in the box. *Courtesy of Judith Miller.* Box without cellophane: $5-12; with cell. intact: $8-15; models still fastened in box: add $5 plus the value of the models.

This "3 Gaited Saddlebred Mare and Foal" (made only in the 1960s) are still rubber-banded in their box. *Courtesy of Shirley Ketchuck.* Box without cellophane: $5-12; with cell. intact: $8-15; models still fastened in box: add $5 plus the value of the models.

Left: The 7", ebony woodcut, "3 Gaited Saddlebred Stallion Mare Foal" box has a horse show scene background. *Boxed models, courtesy of Jaci Bowman.* Box without cellophane: $5-12; with cell. intact: $8-15; models still fastened in box: add $5 plus the value of the models.

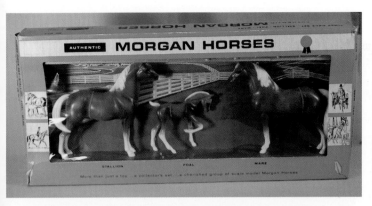

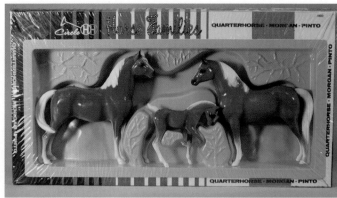

The copper sorrel, 7" Morgan Family by Hartland Plastics, 1960s, came in a dark green box. *Boxed models, courtesy of Jaci Bowman.* Box without cellophane: $5-12; with cell. intact: $8-15; models still fastened in box: add $5 plus the value of the models.

This 7" Morgan set still in its Durant Plastics box is proof that the bright, red-orange Morgans without shadings were from the 1970s. *Boxed models, courtesy of Jaci Bowman.* Box without cellophane: $5-12; cell. intact (never opened): add $5 plus the value of the models.

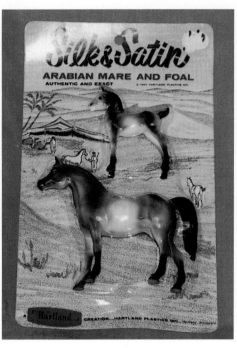

The 1960s, golden-yellow, unpainted, 7" Arabian Mare and Foal were sometimes sold on a card for $1.00 as catalog #6761. *Courtesy of Tina English-Wendt.* Never-opened package with mint models: $25-30.

The 1960s, 5" series Arabian Mare and Foal in painted colors were always sold on a card and called, "Silk & Satin," regardless of color. This set is gray. *Courtesy of Judith Miller.* Never-opened package with mint models: $27-35.

This 1960s, 5" series, bay Arabian set sold for 98 cents. Never-opened package (dated 1961) with these mint models: $22-26.

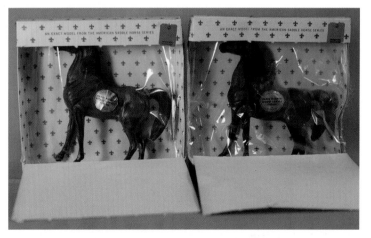

Left: This 1960s package proves that the sorrel, 5" series Arabian Foal was paired with a bay mare! There were no sorrel mares. *Models, courtesy of Jaci Bowman.* Never-opened package with these mint models: $24-28.

The 6" woodcut Arabian Stallions (made only in the 1960s), shown here in walnut (*left*) and cherry, were priced at $1.27. Their sticker reads, "Hand Rubbed/Wood Carved/Finish/by Hartland." *Boxed models, courtesy of Jaci Bowman.* Box without cellophane: $3-8; with cell. intact: $4-10; model still fastened inside: add $5 plus the value of the model.

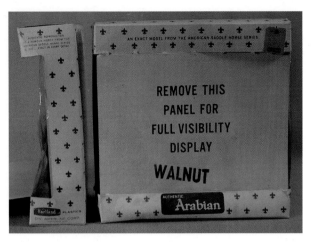

Right: This must be the first style of Tinymite box, from 1965, because it holds the old-mold Quarter Horse, which was black. *Boxed model, courtesy of Shirley Ketchuck.* Box without cellophane: $1-2; with cell. intact: $2-3; then add the value of the model.

These boxes for the 6" woodcut Arabs still have the protective panel that, if detached at the top, can act like a drawbridge. *Boxes, courtesy of Jaci Bowman.* Add $1-2 for the plain, protective panel. Add $3-5 for the protective panel for 1960s 9" series and Family series boxes because they have a large image of the models on them.

Right: This 1960s, Tinymite box, with a new-mold Tennessee Walker, has the price, 29 cents, printed on it. *Courtesy of Pam Young.* Box without cellophane: $1-2; with cell. intact: $2-3; then add the value of the model.

A third type of 1960s Tinymite box has neither the price, nor the empty price circle. This one holds a new-mold Morgan. *Boxed model, courtesy of Sande Schneider.* Box without cellophane: $1-2; with cell. intact: $2-3; then add the value of the model.

Four, Durant Tinymites sold for 87 cents on this red-violet and yellow-green card reading, "1970 Strombecker Circle H Durant, Oklahoma Made in USA Printed in USA." Cardboard backing: $1-2; never opened (with blister-pack cover still sealed): add $5, then add the value of the models.

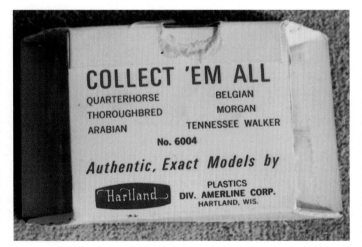

The back of the 1960s, Tinymite box lists the six breeds: Quarter Horse, Thoroughbred, Arabian, Belgian, Morgan, and Tennessee Walker, all #6004. *Box, courtesy of Sande Schneider.*

A black Tinymite Morgan came sealed in this box that reads, "1970 Strombecker, Circle H Durant, Oklahoma Made in U.S.A." *Courtesy of Eleanor Harvey.* Cardboard: $1-2; never opened: add $5 and the value of the model.

Models in this chapter include molds E1-E3.

Mold E1. This 4.75" high white horse on a 1.25" base, a novelty item, has Hartland's pre-1954 "Diamond 'I'" mold mark. *Model, courtesy of Sande Schneider.* $15-35.

In the 1960s, Hartland manufactured the Butazolidin horse for Jen-Sal using the 5" series Arabian Mare mold in clear and white styrene with yellow and white filler. "Bute" is a pain medication for horses. *Courtesy of Judith Miller.* $20-40 with base; $15-35 without base.

Hartland officials were unable to recall this 1960s, Old Milwaukee beer sign, but it includes an apparent Hartland horse. It is not electric. Entire sign (VG-NM): $50-65.

The gold-plated horse in the Old Milwaukee On Tap sign is evidently from the 5" series Quarter Horse Mare mold. Horse alone (VG-NM): $15-20.

Left: Hartland Plastics made this Mustang Malt Liquor wall sign using the 9" woodcut Mustang mold in Tenite, no earlier than 1965. The base under the horse lights up. *Courtesy of Heather M. McCurdy.* Entire sign: $35-65; horse alone: $30-60.

Right: A counter display for Mustang Malt Liquor uses the same, golden Mustang. *Display, courtesy of Michelle Smalling; photo, courtesy of Sande Schneider.* Entire sign with bottle: $35-65; without bottle: $30-60; horse alone: $30-60.

Mold E2. Clydesdales manufactured by Hartland have the right hind foot lifted three-quarters of an inch off the ground. Among those, some have a raised rein loop on the harness collar *(right)*, and others do not.

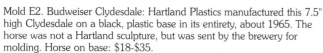

Mold E2. Budweiser Clydesdale: Hartland Plastics manufactured this 7.5" high Clydesdale on a black, plastic base in its entirety, about 1965. The horse was not a Hartland sculpture, but was sent by the brewery for molding. Horse on base: $18-$35.

Mold E2. Two-Point Clydesdale with flat loop—Hartland Plastics made the 7.5" trotting Budweiser Clydesdales with right hind foot three-quarters of an inch off the ground and a flat (non working) rein loop on the harness. $15-$28.

Mold E2. Two-Point Clydesdale with raised loop—I added a rein to demonstrate the working, harness loop at the shoulder and the mouth hole. This 7.5" horse also stands on only two feet; many or all like this were Hartland-produced. $12-$23

Left: Hartland Plastics supplied the 7.5", two-point, trotting Clydesdales for many Budweiser signs and souvenirs. Hartland did not make the signs, but it supplied many of the horses used on them. Sign: $25-40.

Right: The 13.5" square, Budweiser signs with two-point Clydesdales (with raised loops) facing right and left were designed and manufactured by Lakeside Plastics of Minneapolis and Chicago. Sign: $25-40.

Hartland Plastics supplied 7.5" Clydesdales to the companies that made Budweiser signs including this "Big Eight" hitch measuring almost six feet long. *Photo, courtesy of Sandy Bellavia and Kim Fairbrother.* $400-1,000.

Toe-Touch Clydesdale—The 7.5" high, trotting Clydesdales with right hind toe touching the ground are marked "Made in Hong Kong 224"; they were not made by Hartland, but were a brewery souvenir. $8-15.

Flat-Foot Clydesdale—The 7.5" high trotting Clydesdales with both hind feet flat on the ground are marked, "Goldenplum 225 Made in Hong Kong," with an SW logo; Hartland did not make them. $5-12.

Mold E3. Hartland Plastics made this Budweiser sign with horses lacquered gold over silver color. The horses and wagon are 12" long overall; the entire sign is 15" long. Gold or silver hitches in other sizes were not Hartland work, nor were hitches with bay horses 3.5" high. *Photo used with permission.* Entire sign: $35-75; without bottle: $30-60; hitch only: $20-40.

Founders of Hartland Plastics—Iola Walter holds a palomino Large Western Champ at a trade show in Chicago. Edward Walter *(left)* and his wife, Iola, founded Hartland Plastics. Hartland often used Tenite (acetate) plastic, a product of the Tennessee Eastman Chemical division of Eastman Kodak Company. Pete Fox *(right)* a Tennessee Eastman salesman, visited Hartland Plastics regularly. Hartland's Edwin Hulbert thought this photo was taken in June 1951 (or June 1950) at Navy Pier. Edward Walter died in December, 1951. *Courtesy of Paul E. Champion.*

Sculptor Roger Williams always worked at home. His wife, Idella, assisted him. (1950s photo.) *Courtesy of Idella M. Williams.*

About 1955, Hartland Plastics received recognition from State Senator Chester Dempsey *(second from left)*, who is congratulating Charles Caestecker, owner of Hartland Plastics. Also shown are Hartland Plastics executives Robert McGuire *(far left)*, Paul Champion *(third from right)*, and Edwin Hulbert *(far right)*. E. L. Jacobs, a local merchant, is *third from left*. Hartland Plastics was one of the largest employers in its area. *Courtesy of Paul E. Champion.*

Molding press operator Florence Cronce takes horse parts out of the machine and cuts them off the runners in this April 1957 photo. The assembly fixture is at right, and the cooling fixture is at left.

Women package horses with riders at Hartland Plastics, April 1957. The 9" series standing/walking palominos on the table are Roy Roger's Trigger.

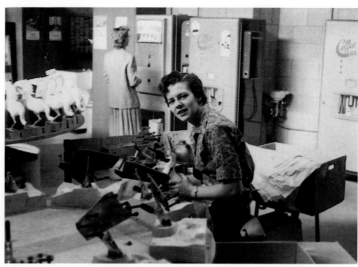

At lower left, a lady soaks halves of horses in a solvent tray and then puts them together. Assembled standing/walking horses from the rider series rest in the cooling wheel so their legs can harden, 1958.

Anita Marquardt (later, Mrs. Jerome Delsman) assembles semi-rearing horses (with mane up and fancy tail), December 1957. Note the other lady at the vending machine and the horses on the cooling wheel *(far left)*.

From left: Richard Schesler, tooling foreman; Hans Seuthe, molding supervisor; Hein Radix, plating supervisor; and Carl Postulart, a machine setup man and later, second-shift foreman, at the company Christmas party, 1957. *(Marks on two faces are damage to the photo.)*

Hein Radix takes the wax master of 5" rider series horse mold cavities out of the copper electroplating bath. There are two left and right sides of the walking horse and one left and right side each of the head down and head up horses. *Reprinted (with permission) from* Modern Plastics, *October 1960, a publication of Chemical Week Associates, Inc.*

Right: Sculptor Alvar Bäckstrand puts finishing touches on an antique car model Hartland Plastics made for a brewery about 1975. This photo with window-pane effect was used in the company's brochure to promote its custom molding services.

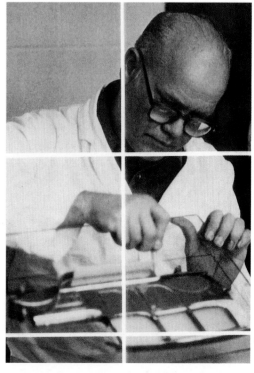

Idella and Roger Williams about 1971. After he retired from Hartland Plastics, Williams' woodcarvings done for fun won awards and appeared on magazine covers. *Courtesy of Idella M. Williams.*

Edgar Schmidt, sculptor Alvar Bäckstrand, John Timmerman, Hans Seuthe, and John Scullin, 1977. Edgar Schmidt made the assembly fixtures for the horses. Hans Seuthe was molding department supervisor from 1949-1977; John Timmerman replaced him after he left. John Scullin was the draftsman engineer who made drawings for the mold parts, including for horses and riders.

Robert and Rita (Mrs. Robert) McGuire posed in 1987 with the four 11" series, 1960s Hartland horses that never went into production (and an unidentified metal bull that is *not* the Schlitz Malt Liquor Bull). Robert McGuire, an attorney and horse lover, was vice president and a marketing executive of Hartland Plastics from 1950-1972. *Photo by Marney Walerius.*

In 1995, Alvar Bäckstrand and Henny (Mrs. Alvar) Bäckstrand posed with models given to him when he retired from Hartland Plastics in 1977. Alvar Bäckstrand sculpted about one-half of all the Hartland figurines from 1956, when he joined the company, until 1969, when model production at Hartland Plastics ended. Some of the models shown here were actually by Hartland's other sculptor, Roger Williams. The Bäckstrands no longer own their collection.

Thomas E. Caestecker, Bernice (Mrs. Paul) Champion, and Paul Champion in 1995. Thomas Caestecker's father, Charles, owned Hartland Plastics during most of the time it made horses and other models. Producing baseball statues was Thomas Caestecker's idea. Paul Champion was national sales manager of Hartland Plastics from 1948-1966.

Idella (Mrs. Roger) Williams in 2000 with a Hartland 11" Arabian her husband modeled. Each year, the National Sculpture Society awards the Roger T. Williams prize for sculpture.

Hartland Plastics' first location, at 132 (now 140) Cottonwood Ave. in Hartland, was demolished, but its second location was this building at 112 W. Capitol Dr. in Hartland, Wisconsin (photographed in 1997). Now subdivided, the building has a two-story section at left and a one-story wing that wraps around the corner.

The Hartland Plastics plant at 112 W. Capitol Dr. was a half-block from this view of Hartland's small, but upscale, downtown.

The original Hartland company's third and final location was this plant at 340 Maple Ave. in Hartland. The Hartland Plastics plant closed in 1978, but its former building has nearly always been occupied since then. (March 1996 photo.)

In the 1960s, there were four horses that got only so far as the metal model phase of mold production, and were never manufactured. One was this standing, 11" series Three-Gaited Saddlebred with roached mane and shaved tail. *Photo by Marney Walerius.*

Model production ended at Hartland Plastics in 1968-1969 before the mold for this 11" series, standing Quarter Horse was finished. *Photo by Marney Walerius.*

This 11" series Tennessee Walker in a big-striding running walk was planned, but never produced in plastic, in the 1960s. *Photo by Marney Walerius.*

The McGuire family still owns the metal models for the four, never-produced 1960s horses, including this 11" series Thoroughbred. *Photo by Marney Walerius.*

(Mr.) Bev W. Taylor owned Steven Manufacturing Co., in Hermann, Missouri, from 1956-1992 and 1995-1997. Steven, the third manufacturer of Hartland models, made horses from late 1983 to November 1994. In this 1998 photo, Mr. Taylor, who is also a magician, autographs *Bev Taylor's Town House Magic* by Bruce Hetzler, Ph.D.

Many Steven horses and baseball players stand loose or in packages in this 1996 factory photo. At upper right are horses in 1990 clam shell packages. Burgundy boxes with blue interiors are from 1993-1994. In the middle, on the floor, are picture frame, through-view boxes used for 9" series horses from 1984-1989 and 11" series closed boxes used from 1983-1986. In 1987-1990, 11" series horses had blue, cellophane-front boxes.

This picture taken during the flood of July 1993 shows the two-story, Steven plant at 224 E. Fourth St., Hermann, Missouri, where Hartland horses were made beginning in 1983. Originally built as a shoe factory around 1900, it has since been torn down. Hartland production resumed after the flood, but ended in November 1994.

The sand bags held for a day and a half, but the retaining wall separating the Steven plant from Frene Creek, off the Missouri River, collapsed at noon on July 8, 1993. The pressure of the flood water knocked out a brick wall of the plant and flooded it and another Steven building nearby. An old sign reads, "Handi-Pac. Inc., Home of Steven's Pixie Toys."

Right: The 8.75" high, trotting Friesian by Kathleen Moody appeared in the 1994 Hartland (Steven) catalog as #251, "Ebony," but the model in the catalog was only wax (painted black), and the mold for the Friesian was never finished. This is the clay original. *Courtesy of Kathleen Moody/Elite Decorations.*

Because of the financial setback of the 1993 flood, seven equine models slated for production in the 1990s were never made, including this walking, Morab stallion by sculptor Kathleen Moody. *Courtesy of Kathleen Moody/Elite Decorations.*

Kathleen Moody's Missouri Fox Trotter sculpture for Hartland depicts an earlier phase of the fox trot than her Fox Trotter eventually issued by Breyer. Hartland never produced its Fox Trotter. Kathleen painted the original, clay Hartland sorrel. *Courtesy of Kathleen Moody/Elite Decorations.*

Equine models usually have ears forward, but Kathleen Moody sculpted #252, "Sassy," the 9" Missouri Mule, with ears attending in two directions. The Mule in the 1994 catalog was wax painted gray; the original, shown here, is unpainted clay. Its mold was never finished. *Courtesy of Kathleen Moody/Elite Decorations.*

Left: This Kathleen Moody, clay original for Hartland (Steven) is a Trakehner doing the piaffe (trotting in place). The 1993 flood canceled Steven's plan to mold it in plastic. *Courtesy of Kathleen Moody/Elite Decorations.*

The gorgeous, Leaping Lipizzan (doing the capriole) by Kathleen Moody was also never molded. In December of 1995, Steven hoped to resume Hartland production in 1997, but it didn't happen. *Courtesy of Kathleen Moody/Elite Decorations.*

Linda Lima sold this 11" scale, trotting, clay Percheron sculpture to Steven (Hartland) in 1988, but due to the flood, it wasn't produced. In 1996, Steven officials reported that the seven, clay equine sculptures had survived the July 1993 flood, but were never seen after the move to the new building in fall 1994. Their whereabouts is unknown. *Courtesy of Linda Lima.*

Carol Gasper *(left)*, Kathleen Moody, and Linda Lima sculpted limited-edition resins that were sold by Paola Groeber under the Hartland name in the late 1980s-early 1990s. Kathleen and Linda also created the seven equines Steven (Hartland) intended to mass-produce in plastic in the 1990s. *Courtesy of Carol Gasper, Linda Lima, and Kathleen Moody/Elite Decorations.*

Since 1997, Breyer, a manufacturer of plastic animals, has issued a Missouri Fox Trotter sculpted by Kathleen Moody. Its ribbons are on the opposite side of the neck from the Fox Trotter she made for Hartland. This bay model is Breyer #768.

Sample and test color models by Paola Groeber, 1987-1990, sold in 12/1994. *Top:* Lady Jewel in unpainted, clear Butyrate plastic (one of four); buckskin pinto Mustang (one of seven); pearled white Polo Pony (unique). *Bottom:* Polo Ponies in Dark Buckskin (one of three) and Dark Dun with shoulder and leg stripes (one of three); metallic copper 9" Foal (one of six); Buckskin 11" Saddlebred (unique). *Courtesy of Paola Groeber.*

Rare models by Paola from 1987-1990, sold by her in 12/1994—*Top:* 11" Quarter Horses in pearl white (*far left and far right*, 12 made), and in colors with silver base coat: Dark Buckskin (six plus one Dark Dun with shoulder and leg stripes), Midnight Blue (three made), Red Bay (five made plus one Brown Bay), and Blue-Black (unique). *Bottom* (all resins): white and bay Miniature horses; Dark Bay Peruvian Paso; and Simply Splendid, the Arabian Stallion, in bright chestnut, sorrel, and pearled white. *Courtesy of Paola Groeber.*

Western saddles and accessories available from Paola included: saddles to fit 7" to 11" size horses, with tan or red suede seats, rounded or square skirts, open or covered stirrups, embossed floral or oak leaf patterns, and matching breast collars and saddlebags in maroon, light tan, and antique walnut. The models are, *from top left:* 9" Thoroughbred in chestnut appaloosa (five made), 7" palomino Quarter Horse Stallion (12 made), Breyer Galiceno Pony, 11" Quarter Horse in grulla; *bottom:* palomino 11" Quarter Horse, and Breyer appaloosa, Stud Spider. *Courtesy of Paola Groeber.*

This new building in the Hermann Industrial Park was built to house Steven Mfg., and has been leased to Steven on a 15-year lease-purchase agreement. After the July 1993 flood, the company was divided between four locations before consolidating here in October 1994. (1996 photo.)

This 1996 view of the Steven plant at 104 Industrial Drive in the Hermann Industrial Park includes loading docks and rolling hills. Hartlands were produced here briefly in 1994. After a change of ownership in 1997, most other Steven toy lines were discontinued. In early 2000, the site was almost a "ghost town."

A 7" bay Arabian Mare and Foal from the 1960s. *Courtesy of Eleanor and Shay Goosens.*

The ten horses of the 9" Breed ("Individual") series are among Hartland Plastics' best horse sculpture. For spirit, refinement, and grace, they are among the finest, mass-produced model horses in any medium. Unlike the Horse & Rider series mounts, these 10 are not generalized horses, but members of particular breeds. The 9" Breed series horses were always sold individually, never with a rider and not in family sets. Thus, they are sometimes called the 9" Individual series. Most, however, are closer to 8" in height than 9".

A c.1963 Hartland catalog described them as:

Authentic Scale Model Horses

A complete new line of the most beautiful, authentic, and exactly decorated individual horses available. No detail has been overlooked in making available a collector's set of large horses that are perfect in anatomy and color. These individual horses are 8.25" high and 9" long. Molded of tough cellulose acetate—they are both beautiful and durable.

A 1964 Hartland catalog added these lines:

The authority on horses will note that our sculptors have caught exactly those physical characteristics which differentiate one breed from another. Every horse lover will certainly want the entire series.

Action Poses. This series of nine adults and a weanling foal includes only two models in standing poses: the Arabian Stallion and Weanling Foal. The other eight depict action that is fluent or powerful, yet accurate and understated. The horses (mold shapes) of the 9" breed series and their heights (in inches) during their three manufacturing eras are listed below. The numbering begins with mold #12 because molds #1-11 were the horses designed for riders.

9" series Model Heights	1960s	1970s	1980s-1990s
12. Thoroughbred	7.5"	—	7.75"
13. Arabian Mare (Grazing)*	<5.75"	—	>5.75"
14. Five-Gaited Saddlebred	8"	—	8.25"
15. Polo Pony	8"	—	8.25"
16. Arabian Stallion	8.75"	—	9"
17. Weanling Foal	6"	—	6"
18. Mustang (Rearing)—Smooth	9"	—	9.25"
19. Mustang (Rearing)—Woodcut	9.25"	>9.25"	—
20. Tennessee Walker (Woodcut)	8"	—	—
a. mane long on left			
b. shorter mane on both sides			
21. Three-Gaited Saddlebred (Woodcut)	8.5"	>8.5"	—

*The height of the Grazing Mare is measured at the highest point: the withers. Symbols: < means just under the height given; > means just over the height given. A dash (—) means that the model was not made in that time period.

Note that the Weanling Foals remained the same height (always about one-eighth of an inch over 6"), but the other 1980s-1990s models are one-quarter of an inch taller than their 1960s counterparts. In the Tennessee Walkers, there are two major mold sub-types: those with the mane on the left, and those with the mane arranged equally on both sides.

Stallions and Mares. The 9" series includes an Arabian Mare and Arabian Stallion. The woodcut Mustang was called a "Wild Stallion" or "Rearing Stallion." The smooth Mustangs and woodcut Tennessee Walkers and Three-Gaiters were also stallions, but Hartland stallions were, technically, geldings. The Thoroughbred, Polo Pony, and Five-Gaiter were mares. The Weanling Foal is a filly. The total is five males and five females.

Began in 1963. Hartland Plastics' 9" Breed/Individual series horses debuted in spring of 1963 and were sold until 1969. The second Hartland company, Durant Plastics, produced only the woodcut Mustang and woodcut Three-Gaiter. Seven others were made by the third and fourth Hartland manufacturers, Steven and Paola, in the 1980s-1990s. The only 9" series horse (mold) that has not been made since the 1960s is the Tennessee Walker, but they are plentiful nonetheless.

The first 9" Individual series horses were the Thoroughbred, Arabian Mare, and Five-Gaited Saddlebred, shown in a Frederick C. Wolf & Son catalog dated April 9, 1963. An undated, Hartland catalog, likely from early 1964, shows those three and mentions that the Polo Pony, Arabian Stallion, smooth Mustang, woodcut Mustang, Tennessee Walker, and Weanling

Foal will be available in March, 1964. The final 9" series horse, the Three-Gaited Saddlebred woodcut, debuted in the 1965 catalog.

Roger Williams, Alvar Bäckstrand. The sculptor for all ten horses was Roger Williams, and sculptor Alvar Bäckstrand readied Williams' metal originals for mold-making.

Woodcut Horses. For the three, woodcut horses, Bäckstrand also created the whittled look. After the metal model was made from the clay model, Bäckstrand used an engraving tool to put carving marks on the metal model, which was then used to make the mold. The originals were not carved from wood, and the models mass produced from the molds are, of course, plastic.

Tenite Plastic. The 1960s, 9" series horses were always made of durable, high quality, cellulose acetate plastic, frequently identified in Hartland catalogs by a brand name, Tenite. In the 1960s, the 9" series horses were never made of styrene (the cheaper plastic); that came later. When the term "Tenite" is used in this book, it refers to cellulose acetate plastic.

Regular, Classic, or DeLuxe. Numbering and series names for the 9" horse series were not consistent during the 1960s. From 1963-1965, 9" series models each had their own, three-digit number. At that time, the Thoroughbreds, Arabian Mares, and Five-Gaiters were usually $1.98 while the Polo Ponies, Arab Stallions, and smooth Mustangs were often $2.49. An exception was that the Arab Mare did not appear in the 1965 catalog. In 1966, the Arabian Mares were still absent from the catalog, but the other five 9" models (and an additional Mustang) were grouped as the #8000 assortment.

In 1967 and 1968, the 9" horses belonged to either the "Classic Series" or the "DeLuxe Series." The Classic series models (#8001) were white with gray trim (mane, tail, etc.) in 1967, but changed to other colors in 1968. Likewise, the DeLuxe series (#8679) changed model colors between the 1967 and 1968 catalogs. In 1968, there was also an #8000 group of models, but they were the same horses as the 1968 #8001 Classic Series, just packaged differently, "in a decorated, self-shipper counter display," instead of "individually boxed in shrink wrapped gift carton." Ones I've seen came in a box with a cellophane front. A point of consistency in 1960s numbering was that the Weanling Foals in all three colors were always #6100.

Priced from 98 Cents to $2.98. In the 1960s, the Weanling Foals sold for 98 cents, and the woodcut horses were $2.98. The rest of the series sold for $1.98-$2.98 between 1963 and 1969, a time of little price change in the U.S. economy. Classic Series models were more plain and sold for $1.98 (later, $2.39). DeLuxe Series models, more elaborately painted, were $2.49 (later, $2.98). Models selling for $2.49 or $2.98 had eye whites painted in (although real horses do not usually show much white of the eyes).

In 1964, the Sears and Spiegel Christmas catalogs each sold 9" Hartlands. The Mustang and Tennessee Walker woodcuts were $2.59 from Sears; $2.47 from Spiegel. Sears sold the dark bay Polo Pony and copper sorrel Five-Gaiter for $1.98 while Speigel sold the Polo Pony and bay Thoroughbred for $1.67 each.

1970s Models. Durant Plastics (of Durant, Oklahoma) made only two 9" series models: the woodcut Mustang and woodcut Three-Gaiter. Durant issued both in an orange-brown plastic similar in weight to the Tenite Hartland Plastics used. Their color corresponds to the walnut (tan) woodcut horses Hartland made. They may look shinier than the 1960s models because less stain was used to antique them. The 1960s models usually have a matte finish or are shiny with no stain at all. The seams may be a little less perfect on the Durant models, but it is not always possible to tell the Durant woodcuts from the Hartlands. Packaging was a box wrapped in cellophane; an entire side of the model was visible. The Durant catalogs did not include the price, but it was probably $2.98.

1980s-1990s Models. All seven of the non-woodcut, 9" series horses returned in 1984 at the hands of Steven Mfg. Co. (of Hermann, Missouri). To make a long story short, in the 1980s-1990s, Hartland horses in the 9" size were styrene, then Tenite, then styrene again, then Tenite again.

The 1984-1986 Steven 9" horses were styrene. Then, from mid-1987 to early 1990, Steven molded them in Tenite for both itself and for Paola Groeber's company (out of her home in New Haven, Missouri).

The next period was two years (March 1990-March 1992) when Steven stopped making 9" series horses in Missouri and sent six of the 9" molds overseas to make horses in styrene for the Three-Stallion and Three-Mare sets in the JCPenney Christmas catalogs in 1991 and 1992, respectively. (No mare sets in styrene have been reported, though, so the supply of Tenite mares in the same colors may have sufficed. My stallion set is styrene, but some collectors reported stallion sets with Tenite Arabs and Mustangs. It makes sense that pre-existing Tenite models were used.) The fourth stage for 1980s-1990s horses was Tenite horses (again) made by Steven Mfg. in Missouri (again) from spring of 1992 to fall of 1994.

During Paola Groeber's transition in 1986 and 1987 from seller of Steven Hartlands to manufacturer of her own Hartland Collectables, she issued test run and sample horses. Many were styrene, but some were Tenite. In December 1994, four years after her company closed, and again in fall of 1999 and spring of 2000, she sold additional test run and sample models in both styrene and acetate. A few of them had been painted after 1990 from her stock of unpainted bodies, but most were made between 1986 and 1990. For test colors only, Paola sometimes included a code for the plastic in her catalog numbers: "T" stood for Tenite, and "H" stood for styrene (HIS, which is Hi-Impact Styrene).

1980s-1990s Boxes. There were three major styles of Steven boxes for 9" horses in the 1980s-1990s. They started in a picture-frame box dated 1984: the cardboard frame wrapped around a clear, rigid plastic holder shaped to the model; it allowed the model to be seen on both sides. Those boxes held the #260-261 Assortment. In 1990, 9" series horses were switched to a clear plastic "clamshell" package that closed at the top. The horses then had names, and the entire group was series #268-600, the "Hartland Champion Collector Series." The 1993-1994 package was a dark red-brown box with cellophane front and blue cardboard backing. The 1991 and 1992 JCPenney sets (Three Stallions and Three Mares) were shipped in plain, cardboard boxes.

From $7 to $16 in the 1980s-1991. The Steven 9" series horses made in styrene from 1984-1986 were sold by various stores and mail-order sellers. Paola Groeber, who was then a seller of Steven models and not yet a manufacturer of Hartland horses, sold the Steven models for $7.15 in 1985, $7.25 in August 1986 and $7.35 or $7.45 in 1987. In 1988, Paola sold her remaining stock of Steven 9" horses at the reduced price of $5.35.

Paola's prices for the 9" series horses she manufactured in Tenite began at $10.50 in 1987, were $10.95 in 1988, and in 1990, her final year of production, reached $13.95, $14.95, or $15.25, depending on the model. Her special runs were $11.50 in 1988 and $15.95 in 1990. The Weanling Foals were $7.65 in 1988, but $8.50 for special runs; in 1990, they were $9.50, $9.95, or $10.25, depending on the color.

In 1990 and 1991 I saw 9" Steven models—styrene models made in 1984-1986 and Tenite models made in 1988-1990—in stores for $13 or $15, depending on the store, not on the type of plastic. The Steven, Three-Stallion set was $24.99, in the Christmas 1991 JCPenney catalog. The following Christmas, the Three-Mare set was $19.99 from JCPenney. In the 1980s, Steven Mfg. was owned by Mr. Bev W. Taylor.

Prices Jumped in 1992-1994. Under new ownership—a group of investors headed by Tim Ford—the 9" series, Steven horses produced in Tenite in 1992-1994 wholesaled for $12 or $13 each and retailed for about $18-$20 from mail-order, model horse sellers. Using a new sales strategy, the 1992-1994 Steven Hartlands were not sold through stores because they were aimed only at collectors, rather than including the toy market. Three, 9" horses were $24 each in the 1994 and 1995 Enchanted Doll House catalog, Manchester, Connecticut. They were the bay Mustang, Desperado; blue roan Five-Gaiter, Twilight Moon; and flea-bit grey Grazing Arabian Mare, Fair Maiden.

Wall Display. In the 1980s, Steven also sold its "#290 Corral Display Frame for Horse Models" from 8" to 11" high. Intended to be mounted on the wall, it had a 16.5" x 13" plastic picture frame and interchangeable background scenes, and a platform for the model to stand on, with a white "corral" fence to surround the model on three sides. The fence style was flat boards with three rails and three posts. The frame, platform, and fence extended a total of 4" out from the wall. It was $8.95 and, later, on special for $6.95. One sold on eBay in 2000 for $27.

1980s-1990s Eye Whites. Some of Paola's 9" series models had eye whites, and some did not; if you saw 10 of the same model, five might have eye whites and five might not. The 1983-1991 Steven horses did not have eye whites painted in, but the 1992-1993 Steven models did.

Mold Marks. To tell Hartland models of the three eras—the 1960s, the 1970s, and the 1980s-1990s—apart, note the color, type of plastic, height, mold mark, and, of course, the model's box (if present). Most of the 1960s horses are marked, "©Hartland Plastics, Inc." The exceptions are 9" woodcut horses, which leave off the "Inc." or have a flattened area where the mold mark would be and, perhaps, a single letter there, instead. Single letters were used to identify molded sections to assemblers, and were not part of the brand identification. Single letters may also be seen on horses marked "©Hartland Plastics, Inc."

The 1970s woodcut horses have, in examples I've seen, just a flattened area.

The 1980s-1990s 9" series horses—the seven, non-woodcut molds—say, "©Hartland," with the words, "Plastics" and "Inc." almost entirely

smoothed off. They may also have a single letter in addition, and the letters can vary. Of my Tenite 1980s-1990s Thoroughbreds, for example, two have a "T" and three have a "B" in addition to "©Hartland." To summarize, a 9" Individual series horse marked "©Hartland Plastics, Inc." is from the 1960s, but one marked "©Hartland" is from the 1980s or 1990s.

1960s Woodcut Alphabet Soup. My three, 1960s woodcut Three-Gaiters have a "Y," but another collector has one with a "D." My woodcut Mustangs and Tennessee Walkers vary. My three, left-mane Walkers say "©Hartland Plastics," but one of my two-sided mane Walkers has a "K" and the other has an "A." (They also have a slightly different whittle pattern, and that pattern is different from the one on the left-mane Walkers.)

Among my woodcut Mustangs, one crisply says, "©Hartland Plastics"; two say, "©Hartland," but the word "Plastics" is so blurred that it almost isn't there; two have only an "S"; and one has only an "X." There are also slight differences in the whittle pattern, with the "X" model different from the "S" models, and both types different from the other three models. Collector Jeannine Bergeron also noticed two different whittle patterns on woodcut Mustangs; it looks like there are at least three types.

Colored Plastic. The woodcuts were molded in black, tan, or dark red-brown plastic. In addition, some 1960s horses in the 9" series were molded in off-white, orange tan, golden-yellow (underlying some Arab Mares painted pearl white), pink (under the paint of a blue roan Polo Pony), maroon, and dark red-brown with a hint of purple in addition to the old standby, white plastic. A very common model, the glossy, shaded maroon bay 9" Thoroughbred is typically molded in maroon plastic. Collector Betty Mertes found very pale green (or pale, robin's egg blue) plastic visible at a rubbed spot on her 1960s, sorrel, 9" Arab Stallion, and a member of her 4-H club has one like it, too.

In the 1980s-1990s, nearly all 9" series models were molded in white plastic and then painted.

Paola sold Unpainted, White Horses. From 1986-1990, white, unpainted, models were actually sold through Paola Groeber's catalogs. Two molds in styrene were advertised, and seven shapes were sold in Tenite. They are listed here, instead of with each horse mold shape. They are illustrated in the Hartland Horses: Details & Oddities section of the photo gallery, rather than with the 9" series photos. Unpainted, white models in Paola's catalogs were:

9" Thoroughbred in white styrene (#231U), available at Christmas 1986 for $4.

9" Grazing Arabian Mare in white styrene (#239U), available at Christmas 1986 for $4.

9" Thoroughbred in white Tenite (#BOD11), 1988-1990.

9" Grazing Arabian Mare in white Tenite (#BOD12), 1988-1990.

9" Five-Gaited Saddlebred in white Tenite (#BOD4), 1988-1990.

9" Polo Pony in white Tenite (#BOD10), 1988-1990.

9" Arabian Stallion in white Tenite (#BOD6), 1988-1990.

9" Rearing Mustang—smooth mold—in white Tenite (#BOD8), 1988-1990.

9" Weanling Foal in white Tenite (#BOD13), 1990, for $5.

In catalogs/price lists from 1988 and June 1989 the six, white Tenite, 9" series horses were $4 each. In September 1990, the price had gone up to $8 each. The Tenite models were described as "suitable for remaking" because they typically had "bent legs, open or mis-matched seams, etc."

Unpainted Brown Polo Pony. A 1988 catalog from Paola sold 30 pieces of an unpainted Polo Pony molded in brown, styrene plastic. It was item #BOD2, priced at $1. It suggests that some of the #236, Seal Brown Steven Polo Ponies from 1984-1986 were molded in brown plastic. I examined two that looked painted brown over white plastic, so there may have been two approaches to the Seal Brown color.

Unpainted, 1960s Models. Except for Paola's white horses and brown Polo Pony and some 1960s woodcuts with no stain, 9" series horses were never sold in an unpainted state. (This applies to the 9" series; in the 1960s-1970s, the smaller horses—7" series and smaller—were commonly sold unpainted, in various colors of plastic.) However, I've seen a 9" Thoroughbred in solid brown, and can only guess that it was removed from the factory before being painted. A collector who has combed yard sales in the Hartland, Wisconsin, area for at least two decades has a Polo Pony molded in dark, red-brown Tenite and a white Five-Gaiter (unpainted). Employees were sometimes allowed to take home factory rejects, which in this case would be unpainted models with a bent leg, gouge, chipped ear, or other damage. Unpainted, 9" models with no apparent flaws may have been pilfered.

The Hartland area has also harbored some partly-painted models that escaped the factory before their paint job was completed. An example is Peggy Howard's metallic, blue roan Polo Pony whose bandages had not yet been painted red; she got it from a Hartland-area collector. Other examples, illustrated in the Details & Oddities section, are useful for seeing how the models were made.

Unfinished, partly painted 9" models are not unique to the 1960s, of course. Unpainted, white models found in the vicinity of Hermann, Missouri, include models that floated away from the Steven factory in the flood of 1993. A Hermann historian, Erin Renn, said that for a while she'd go out in her yard along the river each morning to see what plastic horses had washed up.

Saddlery. In 1989-1990, Paola offered "Junior League" western saddles in black or brown with Arab skirts or square skirts and matching bridle, breast collar, and parade tapederos, to fit 9" or 7" series models and classic-sized Breyers. They were made by Kathleen Bond, who was also making full-sized saddlery for a living. (She said each full-sized saddle took a month to make and sold for $1,000.)

Item Numbers. In the 1960s, 9" breed-series horses were numbered in the 800s or 8000s. The 1970s 9" horses were #4202. In the 1980s and 1990s, Steven horses in all sizes were numbered in the 200s while Paola's were usually numbered in the 400s. Exception to the 200s/400s rule are the 1986 and 1987 test colors painted by Paola Groeber, which are numbered in the 200s and 300s and have a fourth digit, which is a capital letter.

In looking at order forms or catalogs from Paola, note the model number to tell which models she was selling were by Steven and which she painted herself. Paola sold both her own and Steven's models for a while in the 1980s. Her regular 1985, Christmas 1985, and regular 1986 catalogs sold Steven models. In 1987, her regular catalog had two pages: a page of Steven models (with the #270 Indian set at the bottom) and a separate page of her own products. Her 1988 order forms still included some Steven models that had been produced in 1983-1986, but in 1989 and 1990, all the models she was selling were models she had painted herself.

Another thing to remember is that 12 identical colors of 9" models in Tenite plastic were painted by both Steven and Paola; finding a model in Paola's catalog does not exclude the possibility that the actual piece you own was painted by Steven. Steven painted and sold larger quantities of models than Paola did.

Following are the lists of painted 9" series models of all eras, grouped by the ten molds (shapes), from Thoroughbreds to Three-Gaited Saddlebreds.

Mold 12. The 9" series Thoroughbred

The lanky, walking Thoroughbred is found most often in glossy, maroon bay with shadings, a popular model made for five years in the 1960s. The Thoroughbred's height is 7.5" (1960s) to 7.75" (1980s-1990s). Colors are:

1960s, 9" series Thoroughbreds by Hartland Plastics:
• Maroon Bay (Glossy, Shaded, Maroon Bay): Bay, #873 (Tenite), 1963-1967; molded in dark red (brown-red) plastic; black body shadings; black m-t-ll-h; no white anywhere.
• Budweiser Brown (Tenite), 1967-1968, a #8679 DeLuxe model; white plastic covered with orange-tan paint and an overcoat of black shadings; black m-t-ll-h; white (unpainted) face; eye whites. This Thoroughbred's bay color matches the Budweiser Clydesdales Hartland Plastics manufactured in the mid-1960s; that's why the catalog color name is "Budweiser Brown." (It should have been "Budweiser Bay.")
• White-Grey (Tenite), 1967 only, a #8001 Classic model; white plastic; white body with gray m-t-h. Catalogs call this model "white with black trim," and the 1967 dealer catalog portrays one with black m-t-h, but in collector catalogs, the m-t-h are gray, and that is how the model is usually seen.
• Clay Bay: color unnamed in catalog (Tenite), 1968 only, a #8001 Classic model; molded in orange-tan plastic and no paint added to body itself, but m-t-h are black. The 1968 dealer catalog showed the 9" Thoroughbred in pale taupe/oatmeal dun and the 9" Five-Gaiter in clay bay, but for production, the models evidently traded colors, and no Thoroughbreds in pale taupe/oatmeal dun, nor Five-Gaiters in clay bay have ever been reported.

1980s-1990s, 9" series Thoroughbreds by Steven Mfg:
• Brown: Mahogany Bay, #231 (styrene), 1984-1986, molded in white plastic; body painted brown with black points; body color is even (lacks shadings); no white markings.
• Bay Appaloosa, (Tenite), 1992 special of 200 for the West Coast Model Horse Collector's Jamboree banquet in Ontario, California, July 18, 1992, which included 20 raffled Nov. 21, 1992 at the 1992 California Model Horse Championships; matte, reddish bay with spotted blanket pattern.
• Chestnut, #230 (Tenite), "Rapid Delivery," 1993-1994; white plastic under reddish body color; black muzzle; knees, hocks, mane, and tail may have faint black shading, but are mostly reddish; four white stockings and beige hooves.

• Black Dapple Appaloosa, #202-20 (Tenite), 1994 special of 50 for Daphne Macpherson's Cascade Models, which sold 40 and raffled 10 at the 1994 Northwest Congress model horse show.

Colors of 9" series Thoroughbreds by both Steven and Paola:
Note: Both Steven and Paola made "Sandy Bay," but one appears dark buckskin and the other is clear buckskin, so they have separate listings.
• Shaded Buckskin: Sandy Bay, Steven #227 (Tenite), 1988-1990 and part of Three-Mare set by Steven in 1992 JCPenney Christmas catalog; white plastic painted taffy tan with a light, over spray of black shadings on body and face; black m-t-h; knees and hocks are heavily shaded although the lower legs are mainly unshaded; left hind stocking (white); no eye whites. Some of the JCPenney models could be styrene. Some or all of the JCP models have a matte finish while some or all of the regular run examples have a semi-gloss finish, but there is little point in trying to distinguish between them. In the 1990 clamshell package, this model's name was "Caramel Oats" in the series #268-600, "Hartland Champion Collector Series."
• Clear Buckskin: Sandy Bay, Paola #461 (Tenite), 1988-1990, fewer than 1,000; white plastic; body is golden buckskin color with black shading confined mainly to the muzzle and lower legs; black m-t-h; left hind (white) sock; some or all have eye whites.
• Dark Sorrel: Charcoal (Tenite), Steven #228 and Paola #460, 1988-1990; white plastic; body paint looks dark, dull brown, but is technically gray (like the gunmetal gray automobile color); mane and tail are beige; two hind stockings (beige, not white). Models by Paola and Steven look alike. In 1990, this Steven model in the clamshell package was "Charr' d Bars" in the #268-600 "Hartland Champion Collector Series."

9" series Thoroughbreds by Paola Groeber, 1987-1990:
• Dark Dapple Grey, #331A (styrene), May 1987 test run of 50; dapple grey body with dark mane and tail.
• Light Dapple Grey, #331B (styrene), May 1987 test run of 50; dapple grey body with light mane and tail; described in catalog as having "very light shadow dapples."
• Deep, Rich Chestnut, #331C (styrene), May 1987 test run of 100; glossy, brown-red body with black hooves and muzzle; has eye whites; no white markings on most or all; differs from 1993-1994 chestnut Steven model by the different type of plastic and lack of dark shadings on knees and hocks.
• Black, #331D (styrene), May 1987 test run of 18; black with eye whites and one white stocking.
• Light Sorrel (#460L), a variation of Dark Sorrel: Charcoal, #460 (Tenite), about 24-30 made in this color by accident; they are a lighter and warmer brown than the normal Paola (and Steven) Dark Sorrel: Charcoal; beige mane and tail; white plastic; typically, two hind stockings (white) and three black hooves (left hind is pink). The color is raw sienna; it was supposed to be mixed with burnt umber, but Paola's sister, Norma Reed, reportedly made a mistake. The mistake turned out well.
• Metallic Gold Bay, #462 (Tenite), 1989-1990, fewer than 1,000; metallic gold body with black muzzle and black m-t-ll-h; no white markings.
• Dark Dapple Grey, no # (Tenite), show special color dated 4/90 on belly, about 24 made. They have eye whites, a bald face, and black m-t-h.
• Dapple Grey: "Scammer," (Tenite), a 1990 special of 256 for Black Horse Ranch; dapple grey with black mane and hooves; tail is black on upper half, white below; two (white) hind socks. "Scammer" was modeled after a real Black Horse Ranch Thoroughbred and sold with "Black Embers," a 9" series Weanling foal in black, for $28. The models are signed: "H.C. 9/90."
Fewer than 10:
Clear Buckskin: Sandy Bay in styrene plastic; unique because the plastic was normally Tenite, was used for catalog photo, 1989.
Chestnut Appaloosa with spotted blanket and mottling at edges of blanket; five made in the late 1980s.
Blue with white points (Tenite), one with pink hooves was made for a July, 1990, model show in St. Louis.
Copper with white points (Tenite), one made for 1990 St. Louis show.

Mold 13: The 9" series Arabian Mare (Grazing)

This grazing mare, usually referred to as "Arabian" in the 1960s Hartland catalogs, has a pretty, detailed head. Due to the mouth-to-the-ground pose, the model measures about 5.75" high at the highest point, the withers. The 9" series Arabian Mare mold debuted late in 1963 and appeared in 1964, 1967, and 1968 catalogs, but not in the 1965 and 1966 catalogs. It came in six colors in the 1960s, and the most common color is blue-gray with white mane and tail. Five regular colors appeared in the 1980s-1990s, plus special and test colors. A metal copy, 5.25" high and gold in color, has a much thinner tail.

1960s, 9" series Arabian Mares, by Hartland Plastics:

• White with black points (Tenite), late 1963-1964, #876; white body (the color of the plastic) with mane, tail, lower legs, and hooves painted black.

• Blue-Gray (Tenite), 1964, #876 (same number as the model above); dark blue-gray body with black shadings and black lower legs and hooves; the mane and tail are white, the color of the plastic.

• White with gray mane and tail (Tenite), 1967, a #8001 Classic model; white body (the color of the plastic) with m-t-h painted gray; lower legs are white.

• Pearl White (Tenite), 1967, a #8679 DeLuxe model; the mane, tail, etc., of this model match the body; the white paint has the color depth of a pearl; often molded in white plastic, but one molded in golden-yellow is also illustrated.

• Bay with body color molded in (Tenite), 1968, a #8001 Classic model; molded in brown-red (maroon) plastic; m-t-ll-h were painted black; no body shadings; no white on face or legs.

• Bay with body color painted on (Tenite), 1968, a #8679 DeLuxe model; body painted brown-red (maroon) with black shadings over white plastic; black points (m-t-ll-h); no white on face or legs.

1980s-1990s, 9" series Arabian Mares by Steven Mfg:

• Dove Grey: Light Grey, #239 (styrene), 1984-1986; body color is light gray, not white; points are black; plastic underneath is white.

• Buckskin, #241 (styrene),1986-1986; apricot orange body color with black m-t-ll-h. (The catalog showed golden legs, but the production models have black legs.)

• Rusty Grey: Flea-Bitten Grey, #221 (Tenite), 1993-1994; white body with rust-colored mane, tail, knees, hocks, and body shadings; there are also tiny, dark flecks on the body. This model might also be called a chestnut varnish roan, and some collectors call it a strawberry roan.

Colors of 9" series Arabian Mares by both Steven Mfg. and Paola Groeber:

• Light Dapple Grey (Tenite), Steven #249 and Paola #450, 1988-1990; light gray body is speckled with white; mane, knees, and hocks are dark gray; there are four white stockings. The tail is often dark on top, white on the bottom; if the white area ends abruptly, the model is more likely to be by Steven; a smoother color transition is usually a Paola model. Hooves are black (Steven or Paola), two black and two pink (Steven), or three black and one pink (usually Paola). The size of the dapples can vary, with the larger dapples being less common and by Steven. Models by Steven also often have a heavier application of facial shading, but, in general, it can be difficult to tell the Steven and Paola models apart. In Steven's 1990 clamshell package, the light dapple grey Arabian Mare was "Silver Vanity" in series #268-600, the "Hartland Champion Collector Series."

• Chestnut (Tenite), Steven #240 and Paola #451, 1988-1990 and part of Three-Mare set by Steven for 1992 JCPenney Christmas catalog; reddish body with matching mane and tail; four white stockings; hooves are black or three black and one beige; the muzzle, mane, and tail can show slight, dark shadings. Models by Paola and Steven and the JCPenney models all look about alike. (Some or all of the JCP models have a matte finish while the Paola models and some or all of the regular run, Steven examples have a semi-gloss finish, but there is little point in trying to distinguish between them.) In Steven's 1990 clamshell package, this model was "Copper Mist" in the series #268-600, "Hartland Champion Collector Series."

9" series Grazing Arab Mares by Paola Groeber:

• Dark Dapple Grey (Pearled), #339A (styrene), May 1987 test run of 19; silvery, light dapples on a dark gray body with black mane and tail; model has a translucent, pearled finish.

• Light Dapple Grey (Pearled), #339C (styrene), May 1987 test run of 31; silvery, light dapples on a light gray body with dark gray mane, tail, knees, and hocks; model has a translucent, pearled finish. On some models, the mane and tail can appear almost black, but with a pearled sheen.

• Light Bay, #339C (styrene), May 1987 test run of 45; red-brown body with black mane and tail and dark shadings on the knees, hocks, and body; no white on legs or face.

• Light Sorrel: Light Chestnut with white mane and tail, #339D (styrene), May 1987 test run of 45; soft, translucent brown body with white mane and tail; example illustrated has no white on legs.

• Darker Dapple Grey (Tenite); these darker dapple, Tenite models are actually a variation of Paola's #450, Light Dapple Grey. She said that the darker color was made for one year during the 1988-1990 run of this model. The tail is dark on top gracefully blending to white, without an abrupt, white area visible where the model was held by a clamp during painting, as often seen on the Steven #249. No matching (darker dapple) Arabian Stallions in Tenite were made.

• Light Bay, #452 (Tenite), 1989-1990; fewer than 1,000; reddish brown body with black mane, tail, and muzzle; no body shadings; two (white) hind stockings; hooves are typically: three black, one beige.

• Metallic Copper, #453 (Tenite), 1990 only; fewer than 1,000; body painted metallic copper; mane, tail, and lower legs remain white (the color of the plastic); hooves are painted black.

Fewer than 10:

Grey (with dapples) with black hooves, black-and-white m-t; 1989 sample; Tenite, one, sold in 1994.

Grey (with dapples) with one pink hind hoof and four socks, partly-white tail; sample; 1988 or 1989; Tenite, one, sold in 1994.

Rose Grey with dark dapples and black m-t, glossy; signed and dated 1988; Tenite, one, sold in 1999.

Pearl White with pink hooves and muzzle and some dark shading on the muzzle, semi-gloss; signed and dated 3/1990; Tenite, two made, sold in 1999.

Dapple Grey like Paola regular run #450, but in styrene plastic; 1988; one made, used for catalog photo, sold in 1999.

Mold 14. The 9" series Five-Gaiter

The 9" series Five-Gaiter is executing the stepping pace, a slow gait expected of five-gaited Saddlebreds. The model is an example of Hartland sculptor Roger Williams' accuracy in depicting gaits. (For questions on gaits, I recommend *The Horse in Action,* by Henry Wynmalen; *Draw the Horse,* by Paul Brown; *Understanding Your Horse's Lameness,* by Diane E. Turner; and *Horse Behavior,* by George H. Waring.)

Balancing on two flat feet and a toe, this model tips easily. The 1960s, 9" Five-gaiters are often found with more than their share of hip and shoulder rubs from these frequent falls. In 1987-1990, Paola Groeber stretched the left hind foot of each of her models down and forward before painting them. Her models are more stable, but many Five-Gaiters need help standing. The cap from a jar, milk bottle, or camera film container can be the perfect prop, but some need only a nickel to lean on.

The 9" series Five-Gaiter measures 8" (1960s) to 8.25" high (1980s-1990s). The most common color is the 1960s copper sorrel. Colors of the Five-Gaiter are:

9" series Five-Gaiters by Hartland Plastics, in the 1960s:

• Copper Sorrel: Sorrel (Tenite), late 1963-1966, #881; orange body with dark shadings; white m-t-ll-h; blaze on face. Some are "frosted" (more metallic-looking), while others are slightly "deeper" (more orange and more glossy, and less metallic-looking). The deeper ones seem to be molded in cream-colored plastic (perhaps, white plastic that was reground or that yellowed) while the frosted ones, which are more common, typically show up in a whiter plastic. Collectors call this color many things besides copper sorrel, such as: orange metallic, bronze pearl, metallic orange sorrel, and metallic sorrel.

• White-Grey (Tenite), 1967, a #8001 Classic model; white with gray mane, tail, and hooves; some may have black mane, tail, and hooves, instead of gray.

• Pale Taupe/Oatmeal Dun: color not named in catalog (Tenite), 1968, a #8001 Classic model; body is oatmeal or pale taupe, the color of the plastic; the mane, tail, and hooves are painted dark brown. This model is similar to dusty dun, but a true dun horse would also have dark brown lower legs and a dark stripe down the back. The 1968 dealer catalog showed the 9" Five-Gaiter in clay bay and the 9" Thoroughbred in pale taupe/oatmeal dun, but for production, the models evidently traded colors, and no Five-Gaiters in clay bay or Thoroughbreds in pale taupe/oatmeal dun have ever been reported.

• Red Bay (Tenite), 1967-1968, a #8679 DeLuxe model; white plastic; body is painted bright red with black shadings; black mane, tail, and hooves; four white stockings. One without the black body shadings is illustrated, also.

1980s-1990s, 9" Five-Gaiters by Steven Mfg:

• Orange Sorrel: Chestnut, #235 (styrene), 1984-1986; orangy body with some shadings; white m-t-ll-h; the color is not metallic and, compared to the 1960s copper sorrel, this model is a much more plain, muted orange.

• Red Sorrel: Mahogany, #235S (styrene), 1984-1995; in 1984, Steven produced about 200 in this color by mistake; the model was intended to be orange sorrel (#235), but, instead, has a bright maroon body with flaxen (beige) points. Among the red sorrels, about 100 were sold by Paola with a gloss coat added and 100 were matte finish and sold by other Steven distributors.

• Cinnamon: color not named in catalog (Glossy, styrene), 1991, was part of the Three-Stallion set by Steven in the JCPenney Christmas catalog; the model is very common; body appears pale, warm brown (like cinnamon) with darker brown points and dark shadings on the head; there are hind stockings (white). On some, the body color is more solid; on others, more roaned (speckled). Because it was called the Three-Stallion set, this model has "stallion" parts added. The president of Steven Mfg., Mr. Bev W. Taylor, said 20,000 Three-Stallion sets were made.

• Blue Roan, #222 (Tenite),"Twilight Moon," 1993-1994; body is medium gray with black points and nose, two white socks, and a bald face; the body color is solid, not flecked.

• Black Pinto, #222-22 (Tenite), "CJ Kaleidoscope," 1994, a special of 225 for Cheryl Monroe's Great Lakes Model Horse Assn. Expo in Lansing, Michigan, July 2-3, 1994. This model is about equally black and white; hooves are pink; face markings were hand painted and all a bit different. It resembles and was named for a national champion Saddlebred in fine harness and halter; in 1994, CJ Kaleidoscope was 6, owned by Janet Wilde, and at stud in Michigan. Cheryl Monroe said she was told on a factory tour in 1994 that the paint masks were paper and had to be taped on each horse. As a result, "some have more over spray, some have thinner or wider, white areas," she said.

• Black with white dapples, #222-21 (Tenite), "Zest for Living," a special of 200 for Debbie Buckler's all-Hartland model show in Napa, California, July 9, 1994; dark gray (nearly black) body with white, dot-like speckles, white mane and tail, bald (white) face, four white socks, and pink hooves. Of the 200, Debbie Buckler customized 15 of them by painting their m-t-ll-h black; she called the 15 "Starry Night."

Colors of 9" Five-Gaiters made by both Steven and Paola:

Note: Both Steven and Paola made Red Roan, but it is usually possible to tell them apart, so they have separate listings.

• Red Roan (low contrast), Steven #243 (Tenite), 1988-1990; dark brown-red body lightened to almost a cinnamon color due to fine flecks of white paint; the points are dark, brown-red; black hooves on front legs; beige hooves on hind legs; two hind (white) stockings. Compared to the typical red roan by Paola Groeber, there is less contrast between the body and the points; on some examples, it may be difficult to tell the Steven red roan apart from Paola's. In the 1990 clamshell package, Steven's red roan Five-Gaiter was "Scarlett Den" in series 268-600, "Hartland Champion Collector Series."

• Red Roan (high contrast), Paola #432 (Tenite), 1988-1990, fewer than 1,000; accurate color with dark (non-roaned) points and mostly dark head; body color is maroon lightened to almost a grayish pink, with a lot of contrast between the body color and the points and head.

• Light Bay, (Tenite), Steven #244 and Paola #433, 1988-1990; reddish brown body with black mane, tail, and lower legs above white socks on two legs; there are greater or lesser amounts of body shading; some Steven models have much shading, but others don't. Some have a more deeply red body color than others; the known, deeply red models were among the Paola light bays although most of hers match the less deep body color of typical Steven models. In Steven series #268-600, the "Hartland Champion Collector Series" (in the 1990 clamshell package), the bay Five-Gaiter was named "Fieldstone's Drive."

9" Five-Gaiters by Paola Groeber:

• Pearl White with "pink skin" (pink shadings), #235SW (styrene), a Christmas 1986 test run of 11.

• Liver Chestnut, #335 (styrene), a May 1987 test run of 72; white plastic painted rich brown with four stockings and a blaze; paint could be a little sticky; was made to resemble the real Saddlebred "Muscles," (registered name: Hi-Stonewall's Showoff) owned by long-time Hartland collector Tina English-Wendt.

• Dark Dapple Grey, #430 (Tenite), shown in the 1987 and 1990 catalogs, fewer than 1,000; body is dark gray with very small dapples; mane and tail are black; bald (white) face; matte or glossy finish; four black hooves or three black hooves and the right hind hoof is pink/peach.

• Light Dapple Grey, #430, same catalog number as Dark Dapple Grey, (Tenite), shown in the 1988 and 1989 catalogs, fewer than 1,000; body is light gray with very small dapples; mane and tail are black; bald (white) face; three black hooves; right hind hoof is pink.

• Pearled, Blue Roan, #430P (Tenite), 1988, special run: probably 200; body appears gray; points and head are black; lower legs are black, but apparently, some have hind socks; model has a silvery sheen. A variation with a lighter and more blue, rather then gray, body is also illustrated. Paola said that the grayer ones were made before the bluer ones.

• Palomino, #431 (Tenite), 1987-1990, fewer than 1,000; golden-yellow body with white mane and tail, four white stockings, black hooves, and a blaze or stripe on the face. Some of the palominos are more semi-gloss while others are more matte. This is actually a dappled palomino. (Although it is most noticeable on greys, healthy horses of almost any color can show dappling.)

• Pearled White, #431P (Tenite), 1988, special run: probably 200 produced; model painted opaque white with gray shadings on the mane and tail and a black nose and hooves.

• Raven Black, #434 (Tenite), 1990 only, fewer than 1,000; black with slight blue undertone; stripe on face and hind white pasterns (low socks).

• Metallic Gold (Tenite), a 1990-1991 special for model shows: unknown quantity (more than 12, perhaps 50) produced; metallic gold body with white mane and tail, wide blaze, and four white socks, all of which are the white color of the plastic. There are facial shadings, and most had pink hooves, but 12 had black hooves. Originally sold for $25.

• Blue-Black with silver mane and tail (Tenite), Paola painted 25 in 1990, but did not begin to sell them until 1994 (four years after her business had closed).

• White with gray mane and tail and pink nose, a c.1990, special of 12 or 13 for a raffle at a model horse show.

• Wedgewood Blue with pink nose and hooves and bald face; 12 were produced, signed and dated 10/1989.

Fewer than 10:

Pearl Black—#236SP—(styrene)—Christmas 1986 test run of three with random white markings.

Raven Black (not pearled)—#235SB—(styrene)—Christmas 1986 test run of two with random white stockings and blaze.

Palomino—no #—(styrene)—May 1987 test run of three with blaze and four stockings—one is illustrated.

Palomino (matte) with gray hooves and narrow blaze—plastic type unspecified—one made in 1987 and sold 12/94.

Palomino (matte) with brown hooves and narrow blaze—plastic type unspecified—one made in 1987 and sold 12/94.

Dark Dapple Grey with large and dark dapples—bald face—pink nose—four socks—black hooves—plastic type unspecified—one made in 1987 and sold 12/94.

Double-Dappled—like #430 normal dapple grey, but with a double-dappled effect—black m-t, four white socks—one is known, produced during 1987-1990, sold in 1999.

Red Dun with taffy tan paint on body—matching mane and tail—four white socks—styrene—late 1980s—three were made about 1985 and one is illustrated.

Blue Roan Pinto (Pearled), unique, c.1990, sold in 1999.

Red Roan with high white stockings—two made in 1987.

Metallic Silver with white points (Tenite)—one made for St. Louis show, 1990.

Metallic Copper with white points (Tenite)—dated 7/1990—very limited quantity—one made for Tina English-Wendt's 1989 Hartland Lovers of America Show, and another for a show in St. Louis, July 1990.

A metal copy of the 9" Five-Gaiter is metallic gold in color, 7.75" high, and has a tail that streams out behind it. It's sold by vendors of fireplace accessories and decorations.

Mold 15. The 9" series Polo Pony

The Polo Pony is cantering on the right lead. It sports legs wraps and a rolled and bandaged tail. Many of the Polo Pony models have a distinctive, joined-star-and-stripe (white) face marking resembling an arrow pointing upward. Eleanor Harvey and other collectors call it an "arrow star." The Polo Pony is 8" (1960s) to 8.25" (1980s-1990s) high. Colors are:

1960s Polo Ponies, by Hartland Plastics:

• Dark Bay with white bandages: (Tenite), 1964-1966, #883; a DeLuxe model; molded in purplish, dark red plastic and then shaded with so much black paint that the dark red, highlight areas are small and subtle; black points; arrow star on face; eye whites. Collector Sandy Tomezik said the 1960s dark bays came in two shades, one slightly redder than the other, but both quite solid.

• White with white bandages (Tenite), 1967 only, a #8001 Classic model; white (unpainted) body with gray points. The white leg wraps contrast with the gray lower legs. Collector Laura Pervier pointed out that some look white while others have slightly ivory plastic.

• Metallic Blue with red bandages (Tenite), 1967 only, a #8679 DeLuxe model, molded in pink or other colors of plastic; painted metallic blue with black shadings and black points.

• Buckskin with red bandages (Tenite), only in a late 1967 or early 1968 catalog, a #8679 DeLuxe model; white plastic painted drab yellow with brown shadings; black points; muzzle is not dark (it matches the body).

1980s-1990s Polo Ponies by Steven Mfg:

• Seal Brown with white bandages, #236, 1984-1986; molded in white styrene (some may be molded in brown styrene); has black points; the body color is even (unshaded). The model has the same arrow star on the face as the 1960s dark bay, but the 1960s bay was the heavier plastic, Tenite.

• Black with red bandages, #236S (Glossy, styrene), 1984-1986, a special of 1,000; 500 were sold by Black Horse Ranch and 500 were sold by Paola Groeber.

• Black Appaloosa, #219-20 (Tenite); a special of 200 for Debbie Buckler's 1993 all-Hartland model horse show in Napa, California; this black appaloosa has the same spot pattern as the Red Roan ("Leopard") Appaloosa mass-produced model. Debbie reported that, within this group, 16 pieces were dark, brown-black appaloosas, while the rest were black. Bandages were red. On about 10 of them, signed, "Deb" or "BD," she re-painted the wraps to purple.

• Creamy Dun with royal blue bandages, #219 (Tenite), "Quick Twister," 1993-1994; molded in off-white plastic; faint peach body shadings; mane, tail, knees, hocks, hooves, ear tops (the upper half of the ears), and muzzle are black or very dark brown; the pasterns on some are russet brown; wide, faint dorsal stripe; white arrow star on face; eye whites. Collectors sometimes call this color: apricot dun or rose grey. (Also, artist Debbie Buckler sold about 10 that she had "color enhanced" by adding dark shadings and changing the wraps to light blue, instead of royal blue; they are singed "Deb" or DB.")

Fewer than 10:

"Flea-bitten Red," donated to 1993 Jamboree auction—one.

Liver Chestnut, a sample for a 1994 special run that was never made for an east coast riding club—one.

Colors of Polo Ponies by both Steven Mfg. and Paola Groeber:

• Blue Roan with white bandages (Tenite), Steven #229 and Paola #470, 1988-1990, and part of Three-Mare set by Steven in the 1992 JCPenney Christmas catalog; light gray body with black points; much of the head is black; this model is the true color of blue roan horses. Models by Steven and Paola appear the same. Note, however, that some of Paola's blue roans are noticeably darker gray than her others and the Steven blue roans. In 1990, the blue roan Polo Pony in the Steven, clamshell package was "Poco Go" in the #268-600 "Hartland Champion Collector Series." Some of the JCPenney models could be styrene.

• Red Roan Appaloosa with black bandages (Tenite), 1988-1990, Steven #228 and Paola #471 (both companies called it "Leopard Appaloosa"); white plastic; white and pinkish body with dark red spots concentrated on a white hip blanket; black hooves and muzzle shading; body appears pinkish because white plastic shows through dark red paint in a pattern of tiny, white spots. Models with heavy, black facial shadings are more likely to be by Steven; models with fainter facial shading and no black on the knees and hocks are more likely to be Paola's work. Overall, models by Steven and Paola appear about the same. Steven's 1990 name for this model was "Joker"; in the clamshell package, it was part of the #268-600 "Hartland Champion Collector Series."

Polo Ponies by Paola Groeber:

• Light Dapple Grey with black bandages, #236A (styrene), Christmas 1986 test run of 13.

• Rich, Dark Bay: "Blood Bay with white bandages" (styrene), a May 1987 test run of 92. The catalog pictures, in black-and-white, a fairly dark bay with white on the face that is less than a bald face although the sales list said, "full face blaze." Using the example of Eleanor Harvey's model, the m-t-ll-h are black and there is an arrow star on the face; this dark bay color is lighter than the 1960s dark bay and shades to black on the muzzle; the bandages were airbrushed, but not masked, so some black paint is on the white wraps. In contrast, the Steven #236 Seal Brown regular run, does not have shadings on the body or muzzle.

• Bay Blanket Appaloosa with white bandages, #336A (styrene), May 1987 test run of 60; brown body with brown spots on white hip blanket and white mottling in brown areas on hindquarters; black m-t-h and muzzle.

• Golden Palomino with red bandages (Tenite), a 1989 special run of 100 by Paola (#217S); white plastic; golden yellow body with white mane and tail; black hooves and muzzle. They are signed, "9/89/H.C." inside a heart shape.

• Bright Chestnut with white bandages (Tenite), a 1989 special run of 100 by Paola (#218S); white plastic; brown-red model with white arrow star on face; some models are slightly less red, and slightly more brown, instead.

Note: the palomino and bright chestnut Polo Ponies sold in the Your Horse Source catalogs from 1989-1990 were by Paola, according to her.

Fewer than 10:

"[Light] Bay with white bandages" (Tenite), Christmas 1986, #236A; the catalog said seven, but Paola later corrected the quantity to three. The model illustrated in black and white is evidently a very light bay, and has a bald face.

Dark Bay, Tenite, by Paola, c.1986-1987; its white bandages were hand-painted and not masked; it has an arrow star on the face; owned by Eleanor & Shay Goosens.

Unpainted, transparent Tenite plastic—#236A—Christmas 1986—three.

Copper with red wraps and white mane and tail, Tenite—one made for 1990 St. Louis show.

Blue with white mane and tail and white wraps, Tenite—one for 1990 St. Louis show.

Paola made the next four colors about 1990, but sold them in December 1994:

Dark Dun with dark blue bandages, dorsal stripe, shoulder and leg stripes, Tenite—three.

Dark Buckskin with dark blue bandages, dorsal stripe, eye whites, and varied shading, Tenite—three.

Chestnut with wide blaze and white bandages, plastic unspecified, probably Tenite—two.

Pearl White with deep blue bandages; some metallic flecks in the paint; no eye whites; plastic unspecified, probably Tenite, no dark shading on muzzle—one.

Paola made the next three in the late 1980s or in 1990, but did not sell them until 1999:

Blue Roan, like the normal, Paola #470 blue roan, but in styrene plastic, used for catalog photo, two made.

Pearl White with Wedgwood blue bandages and pink nose, ears, and hooves, and "subtle shading around the nose and eyes," which are tri-colored; Tenite—two made, one signed and dated 6/1989 was sold.

Pearl White with metallic gold bandages and pink ears, nose, and hooves, Tenite—two.

Mold 16. The 9" series Arabian Stallion

One of Hartland's most noteworthy standing models is, as the catalog says, the "graceful and proud" Arabian Stallion. Not many model horses capture the essence of both wind and fire, but this Arabian does.

The most common 1960s 9" Arab Stallion is the (flaxen) sorrel in Tenite; in the 1980s, Steven Mfg. made a similar (flaxen) sorrel, but in styrene plastic. Dapple grey was not a Hartland color until the 1980s, but in the 1980s and 1990s, there were six, 9" Arab stallion models in dapple grey. A very common dapple grey 9" Arab Stallion is the JCPenney special with black knees, hocks, and face and yellow glue showing at the seams. The 1980s-1990s models are 9" high versus the 8.75" tall 1960s Arabian Stallions.

1960s, 9" series Arabian Stallions, by Hartland Plastics:

• Sorrel: color not named in catalog (Tenite), 1964-1966, #884; warm, medium brown color over white plastic; flaxen (beige) mane, tail, and lower legs; black hooves; smooth but not highly glossy finish; has eye whites. Some were molded in plastic that, in rubbed areas, now appears pale green.

• Copper Sorrel: Sorrel with flaxen mane and tail (Tenite), 1967-1968, a #8679 DeLuxe model; metallic, coppery, orange body with flaxen (beige) m-t-ll; has black hooves and eye whites. The body color is metallic or pearled. The plastic underneath was orange-tan on some, but may be white on others.

• White-Grey (Tenite) with light gray, dark gray, or black mane and tail, 1967, a #8001 Classic model; white body; m-t-h are black or, more often, gray. The white, 9" Arabian Stallion with dark gray or black "trim" appeared in the 1967 dealer catalog, but the model pictured in two, Hartland consumer catalogs had gray "trim."

• Clay Bay (Tenite); this model was not mentioned or shown in catalogs, but I think it was produced in 1968 as a #8000/8001 Classic model. The 1968 dealer catalog shows this mold in medium gray with black m-t-h, but no models that color have been reported. The production color was probably clay bay, instead. The model is molded in glossy, orange-tan plastic; body is unpainted and unshaded; the m-t-h are painted black.

1980s-1990s 9" Arabian Stallions by Steven Mfg:

• Bright Sorrel (Glossy, styrene), 1984-1986, Steven #237. This medium-to-dark sorrel with flaxen mane and tail is similar in color to the 1960s sorrel 9" Arabian stallion. However, the newer model is slightly darker, shinier, and taller, and is styrene, rather than Tenite.

• Black (Glossy, styrene), 650 produced—a Steven special at Christmas 1985 of 150 made for and sold by Paola for $15 each (139 were sold separately as #237B and 11 more were sold with a matching foal for an additional $10) plus 500 Steven made in 1986 for Black Horse Ranch; solid black with no white markings, but they have eye whites. Those sold by Paola are signed by her (even though Steven, not she, painted them).

• Dark Chestnut (Glossy, styrene), 640 produced—a Christmas 1985 Steven special of 140 made for and sold by Paola (as #237C) for $15 each, and 500 Steven made in 1986 for Black Horse Ranch; dark, reddish brown with matching mane and tail; random white markings (meaning that white markings on the face and/or lower legs can vary from model to model). There are white hooves below white leg markings. The socks on the 500 BHR models were hand painted white while socks for the 140 Paola sold in 1985 had masked-off white areas.

• Dark Bay (Glossy, styrene), 32 produced as a Christmas 1985 Steven special for Paola Groeber; random white markings; same body color as the glossy dark chestnut model, but mane and tail were hand-painted black by Steven. Of the 32, 10 or 11 were sold with a matching foal. Paola's flyer called the stallion and foal "dark chestnut," but the set she displayed on the

Internet in 1999 was bay, meaning that the body is brown and the mane and tail are black.

- Dapple Grey (Glossy, styrene), Steven special for 1991 JCPenney Christmas catalog: 20,000 produced as part of the "Three-Stallion" set; light gray body with white speckles (dapples); dark gray to black on knees, hocks, and face; mane is medium gray; tail is dark gray on top, lighter on lower half; shadings may vary. Glue at the seams is visible as a yellow line down the back and elsewhere.
- Raven Black (Tenite), 1993-1994, Steven #224, "Desert Prince;" blue-black body with four white socks, black hooves, white on face and white highlight dot on the eyes; paint can get sticky when hot.

9" Arabian Stallions by both Steven and Paola:

- Chestnut (Tenite), 1988-1990, Steven #245 and Paola #421; a rich, medium reddish chestnut with matching mane and tail, hind socks, beige hooves in back, black hooves in front, and a white marking on the lower half of the face. Models by Paola and Steven look about alike. In the 1990 clamshell package, as part of Steven's #268-600 "Hartland Champion Collector Series," this model was "Regal Air."
- Light Dapple Grey (Tenite), 1988-1990, Steven #246 and Paola #422; light gray body with white dapples (speckles); light gray shadings on knees and hocks, muzzle, mane and tail; tail can be dark on top, lighter below. Shadings on models by Paola can be more subtle, but it can be difficult to tell her models from Steven's without confirmed examples placed side-by-side. This Steven model was "Silver Wind" in the #268-600, "Hartland Champion Collector Series" in the clamshell package in 1990.

9" Arabian Stallions by Paola:

- Dark Dapple Grey (Pearled, styrene), Paola #237S, Christmas 1986 test color of 58; very dark gray, almost black, with silvery dapples, black hooves, and lighter lower legs.
- Light Dapple Grey (Pearled, styrene), Paola #237S (same # as the Dark Dapple), Christmas 1986 test color of 35; light gray with silvery dappling; light to medium gray mane and tail; dark shading on face, knees, and hocks. Unlike the JCPenney dapple grey styrene 9" Arab, there is no yellow glue showing at the seams, and the tail is entirely dark.
- Dapple Grey (Pearled, styrene), Paola #337B, May 1987 test color: 25 produced; similar to Paola's 1986 Light Dapple Grey, but the mane, tail, and hooves are darker, there is no shading on the knees and hocks, and there is less dark shading on the face—the muzzle is dark, but the shading does not extend to more of the face; color may vary somewhat from model to model.
- Light Sorrel: Light Chestnut (styrene), Paola #337A, May 1987 test color: 35 produced; body painted warm, translucent light brown; mane and tail are white; lower legs are brown.
- Light Bay (Tenite), Paola #420, 1988-1990; body is a clear, medium red-brown; mane and tail are black; white stripe on face; hooves are black and there is black shading on the muzzle and lower legs except for a left hind stocking; the left hind hoof is beige.
- Pearled White (Tenite), Paola #420P, a 1988 special: probably 200 produced; painted an opaque white color with black muzzle and hooves and gray shadings on the mane and tail.
- Pearled Rose Grey (Tenite), Paola #421P, a 1988 special: probably 200 produced; body is maroon muted to almost a pinkish gray; mane and tail are dark red; mine has a white stripe on face, but no white on the legs.
- Dapple Palomino: Light Chestnut (Tenite), a February 1990, Paola special for a West Coast model horse show; 20 or 30 made; a lighter, golden-yellow body with dapples; white points; black hooves; dark shading on muzzle; has eye whites. Also, one dated 6/90 was donated to Nancy Ellis and Peggy Howard's model show in Newport News, Virginia, in 1990; it has black hooves and nose and very well defined eye whites. Paola said that fewer were made of the dappled palomino than the solid palomino, but she declined to estimate the number.
- Palomino (not dappled): Light Chestnut (Tenite), Paola show special, some are dated September 1989 and others are dated February 1990 or March 1990; quantity unknown, but likely to be 50 or more; deep and even, golden-yellow body with four white socks and four black hooves; black shading on muzzle; no dapples. The one I have has eye whites. (Since purebred Arabs aren't palomino or palomino Arabs aren't allowed into the purebred Arab registry—the model was sometimes described as "golden chestnut with light mane/tail.")

Fewer than 10:

Brown Bay—Tenite—pictured in Paola's 1987 catalog, but Light Bay was the color actually produced—a handful of Brown Bays exist—body color is plain brown, not reddish—it has a left hind stocking.

Rich Bay—Tenite—two hind stockings—pictured in Paola's 1987 catalog, but Standard Chestnut was the color actually produced—a handful of Rich Bays exist.

Pearled Dark Bay (styrene) with no white—a 1984 Paola sample sold in December 1994—unique.

Dark Palomino (Tenite)—a 10/89 sample that turned out darker than the palomino ("light chestnut") show special color—two were sold in 1999-2000.

Rose Grey with dark dapples (Tenite) and black mane and tail; glossy, unique, a sample produced in 1988 and sold in 1999; note that the normal Rose Grey model did not have dapples.

Metallic Copper with white points (Tenite)—one given as a prize at a show in St. Louis, 1990.

Metallic Gold with black mane and tail and white socks (Tenite)—one was given as a prize at the July 1990 St. Louis show.

Mold 17. The 9" series (Weanling) Foal

Hartland Plastics made the "cutest, wobbly-kneed weanling foals you have ever seen," in three colors from 1964-1968. The 1980s-1990s added many more colors of this alertly standing, 6" foal.

1960s, 9" Foals by Hartland Plastics:

- Maroon Bay: Blood Bay (Tenite), #6100, 1964-1968; glossy; shaded maroon bay with black m-t-ll-h and no white; molded in dark, brownish red plastic; matches the maroon bay 9" Thoroughbred, but some are molded in white plastic.
- Palomino (Tenite), #6100, 1964-1968; glossy; body painted translucent yellow over white plastic; white (unpainted) points; some are a little more "golden" yellow than others.
- Sorrel (Tenite), #6100, 1964-1968; glossy; warm brown body with darker shadings and flaxen (beige) mane, tail, and lower legs; black hooves; molded in white or warm brown plastic.

1980s-1990s, 9" Foals by Steven Mfg:

- Black (Glossy, styrene), #300H, Christmas 1985 special of 10 or 11 sold for $10 each by Paola Groeber; signed by Paola Groeber even though Steven, not she, painted it; there are no white markings. It was sold only as a set with a matching, 9" Arab Stallion.
- Dark Bay (Glossy, styrene), #300H, Christmas 1985 special of 11 sold for $10 each by Paola Groeber; this model was a Steven Mfg. product; it had random white stockings. Paola's flyer described the model as "dark chestnut," but a set she displayed on the Internet in 1999 was bay, meaning that it had a brown body and black mane and tail. In the flyer, the model appeared as a solid, dark image so that it was impossible to tell whether the mane/tail were black or brown.
- "Bay Roan" (Tenite), Steven, 1994 special run for the West Coast Model Horse Collector's Jamboree in Ontario, California, August 1994; about 400 or 500 were made; body color is translucent, rusty orange; mane, tail, knees, and hocks are black; four white stockings.

9" Foals by Paola Groeber:

- Light Dapple Grey (Tenite), Paola #490, 1988-1990.
- Pearl White (Tenite), Paola #490P, 1988 special run color.
- Light Bay (Tenite), Paola #491, 1988-1990.
- Chestnut (Tenite), Paola #492, 1988-1990.
- Red Roan Appaloosa (Tenite), Paola #493, 1988-1990.
- Light Bay Appaloosa (Tenite), Paola #494, 1988-1990.
- Metallic Blue Roan (Tenite), Paola #495, a 1989-1990 special; has black points and low hind (white) socks.
- Black (Tenite), "Black Embers," Paola, September 1990 special run of 256 for Black Horse Ranch; sold as a set with dapple grey Thoroughbred Mare, "Scammer." Some are dark gray over black paint, instead of painted solid black. Black Embers has a star on its forehead but no eye whites. There is a left hind sock and pale hoof under it. Sadly, the real BHR Scammer died young.
- Blue-Black with silver mane and tail (Tenite); by Paola, an unadvertised show special painted by Paola in 1990, but most were not sold until December 1994; has eye whites and silver hooves; 30 were made.
- Wedgewood Blue (Tenite) with pink nose and hooves and white points; a 1990 Paola sample run of 12 dated 6/1990; bald face; white areas are unpainted; the blue color is not metallic. The entire front of the muzzle is light-colored with some dark shadings on the sides of the muzzle. The stockings reach relatively high: to the lower part of the knees and hocks.

Fewer than 10 (all Tenite):

Wedgewood Blue with black muzzle and white points—one was given at the July 1990, St. Louis show; the white stockings extend only about mid-way up the cannons, not so high as on the run of 12 blue foals with pink noses.

Brighter Metallic Blue than #495 with black points and low, hind socks (Tenite); this is a brighter blue than the metallic blue roan color; in 1999, Paola sold one dated 6/90; she thought it was a show special color in 1990, in which case there might be more than 10.

Light Bay—1989 sample with higher black socks than the usual #491 model—one known—sold 12/94.

Light Bay with lighter body color than the usual #491 model; one was sold in 1999.

Pearl White with pink muzzle, hooves, and ears—two made—one sold in 1999; signed and dated 3/1990.

Metallic Copper, c. 1990 sample—white mane and tail, random white socks and random white hooves—six were sold 12/1994—they are signed.

Metallic Gold with white points—six made, signed and dated 3/1990; white on the face extends over the front of the muzzle; the white stockings are very high and were masked off (they go almost straight across).

Metallic Gold with white points—one was made for the 1990 St. Louis model horse show; muzzle is mostly dark; the white on the face is narrow (not visible in profile); the stockings, which go midway up the cannon bone, were not masked, so the gold color blends into the white.

Metallic Gold Bay—very limited quantity—one was made for 1990 St. Louis show.

Mold 18. The 9" series Rearing Mustang (with smooth surface, not the woodcut version)

I classify the Rearing Mustang with smooth surface as a different mold than the woodcut rearing mustang with whittled surface and woodgrain marks. Both 9" series horses are posed identically: striking out with the left foreleg, but the smooth Mustang is one-quarter of an inch shorter because its position is slightly less vertical. The smooth Mustangs, listed below, were made in the 1960s and 1980s-1990s.

Three pinto patterns are found among them. The 1960s Mustangs had six white spots on the body: on the neck, shoulder, and hip on both sides. The second pattern, which is missing the neck and hip spots from the horse's right side, is found on some early-1984 Steven styrene black pintos and on the 1991 Steven styrene brown pinto for JCPenney. Most of the 1980s-1990s pinto Mustangs display a third pattern, which omits the left shoulder spot and right hip spot that the 1960s model had.

The height of the Hartland, smooth Mustangs is: 9" (1960s) to 9.25" high (1980s-1990s). Colors are:

1960s Mustangs with smooth surface, by Hartland Plastics:
• Black Pinto (Tenite), #885, 1964-1966: body is about half black and half white in large patches; white blaze extends over the mouth; four black stockings; white belly; eye whites.
• Palomino (Tenite), #889, 1965-1966; body painted translucent yellow; white (unpainted) m-t-ll-h and eye whites.
• White (Tenite) with gray or black mane, tail, and hooves, 1967, a #8001 Classic model.
• Metallic Gold Buckskin: Golden with black trim (Tenite), 1967, a #8679 DeLuxe model; body painted metallic gold over white plastic; black m-t-ll-h.

1980s-1990s Mustangs by Steven Mfg:
• Black Pinto (styrene), Steven #238, 1984-1986. Typically, this model is missing the left shoulder spot and right hip spot that the 1960s black pinto had. An early 1984 variation of this model is missing the right neck and hip spots, but has left-side markings nearly identical to the 1960s model.
• Dapple Grey (styrene) with white points, Steven #242, 1985-1986; body is a medium-to-dark gray with small and subtle white speckles (dapples); dark muzzle; no eye whites. Do not mistake this model for a Hartland copy made in Hong Kong in a similar color.
• Brown Pinto (Glossy, styrene), Steven special for the 1991 JCPenney Christmas catalog "Three-Stallion" set; 20,000 made; same pattern as early 1984 Steven black pintos; has black hooves and black shadings on face.
• Sooty Dun (Tenite), Steven #220-20, special for the August 1993, Model Horse Collector's Jamboree; available with a $40 adult or $35 child's ticket to the Jamboree's ice cream social along with special Breyer and Hagen-Renaker models. The sooty dun Mustangs were delayed by the Steven flood of 1993, so Jamboree hostess Sheryl Leisure mailed them to ticket holders in fall of 1993. The model is charcoal gray with a hint of beige; Steven painter Tina Strubberg said its official color name was "sooty dun." Some collectors call it "smoky dun."
• Red Bay (Tenite), Steven #220, "Desperado," 1993-1994; reddish brown body with black m-t-and legs above four white socks and four pink hooves; bald face; the muzzle is black.
Fewer than 10: Metallic Silver (Tenite)—two made—intended as a Steven special for Cascade Models/1994 Northwest Congress model horse show, but the special was never produced.

Mustangs made by both Steven and Paola:
Note: The Red Dun Mustangs by Steven and Paola look different, so have separate listings.
• Chestnut Dun: Red Dun (Tenite), Steven #247, 1988-1990; body is painted tan and the mane and tail match; four white stockings, bald face, eye whites, black hooves, and narrow brown stripe down the back; glossy finish. This model's color approaches chestnut. In 1990, when sold in the

clamshell package, this model was, "Flame's Dawn" in the #268-600, "Hartland Champion Collector Series."
• Red Dun (Tenite) Paola #440, 1988-1990; muted orange body with dark red points and dorsal stripe; glossy finish. Unlike the Steven red dun, there is much contrast between the body color and points color.
• Brown Pinto (Tenite), Steven #248 and Paola #441, 1988-1990; has the typical 1980s pinto pattern; models by Steven and Paola look about alike. The Steven, brown pinto in Tenite, was named "Bold Review" when sold in the 1990, clamshell package as part of the #268-600 "Hartland Champion Collector Series."

Mustangs by Paola Groeber:
• Black (Tenite), Paola #215S, a 1989 special of 100; blue-black body, blaze, eye whites, and right hind sock.
• Dark Dapple Grey (Tenite), Paola #216S, a 1989 special of 100; black points, small dapples.
Note: Paola said that the Black and Dapple Grey Mustangs sold through Your Horse Source (the 1989-90, 1990, and 1990 supplement catalogs) were hers, not Steven's.
• Blue Roan Pinto (Tenite) Paola #442, 1990-1991, approx. 200 were made; gray-blue and white body; same pinto pattern as typical 1980s-1990s Mustangs; eye whites.
Fewer than 10:
Taffy Sorrel or Deep Palomino (styrene), with white points, one made, 1986; this is the first model Paola Groeber ever painted.
All of the Mustangs that follow are Tenite:
Buckskin pinto with pink muzzle and left forefoot, pink belly, and eye whites—seven made—sold in 1994.
Red Dun; differs from the normal #440 red dun by having dark red lower legs (instead of white stockings), by not having horizontal stripes on the legs, and by having a matte, rather than glossy, finish; one was sold by Paola, 1999.
Blue Roan Pinto in much lighter blue, 1990, one, sold 1999.
Blue Roan Pinto in very dark blue, has eye whites, two sold in 1994.
Blue Roan Pinto in blue-black, almost black, has eye whites, 1989 sample sold in 1999.
Buckskin (beige-yellow body with black points), has eye whites, one known, signed and dated 8/1990, sold 1999.
Pearl White with metallic gold mane and tail and pink nose and hooves—one only—1988 sample sold in 1999.
Metallic Gold with black points; Paola thought she made only one, 1990; sold in 1999.
For the next three colors, one of each was given as a prize at a show in St. Louis in 1990:
Metallic Gold Pinto with white points.
Metallic Copper Pinto with white points.
Metallic Silver with white points (solid-colored, not a pinto).

Mustang Copies. Fairly common copies of the Hartland Mustang are 8.5" high, styrene, and mold-marked: "'P' in a diamond, 654, Made in Hong Kong." I have them in charcoal gray with white points, and in bay. Another copy is 7.5" high, and a pair of them came attached to an 11" x 6" plastic, lamp base that is marked, "Made by Gilbert Products" on the bottom of the base. One horse is white with black mane and tail and the other is dark brown. In the 1982 Delux Saddlery Importers gift catalog, Baltimore, Maryland, this same "Rearing Stallions Lamp," 16" high overall, with a 12" wide shade, sold for $19.95. A mini copy of the Hartland Mustang is only 3" high, gold-colored, and stamped "Hong Kong."

Mold 19. The 9" series Mustang woodcut

Hartland catalogs called this textured model the "Rearing Stallion" or "Wild Stallion." However, many collectors call both the smooth and textured, rearing horses "Mustangs," so I will also. This Mustang looks whittled, so it's a woodcut, but it also shows a simulated, wood grain, so Hartland called it an "engraved woodgrain." I'm calling it the Mustang woodcut or woodcut Mustang.

Hartland Plastics made the woodcut Mustang in three colors: walnut, cherry, and ebony. The second Hartland company, Durant Plastics, issued it from 1970-1973, only in "light antique walnut." It has not been made since, but the woodcut Mustangs turn up often on the secondary market. A typical description for a woodcut Mustang would read, "Hartland Cherry Mustang" or "Hartland Cherry Woodgrain Mustang." One second-hand seller's colorful description was, "Hartland Cherrywood Mustang." These 1960s-1970s models are 9.25" tall. Some are mold-marked with the Hartland name, and some aren't. (See "1960s Woodcut Alphabet Soup" in the introduction to the 9" series chapter.)

Catalog colors of woodcut Mustangs are:
• Cherry (Tenite), #886C, 1964-1966; appears dark brown due to black stain rubbed over red-brown plastic; matte finish.

• Ebony (Tenite), #886E, 1964-1966; black with faint, grayish highlights; matte finish.
• Walnut (Tenite), #886W, 1964-1968; molded in tan color with dark stain rubbed into the crevices; matte finish. The 1967 and 1968 catalogs included only Mustangs in "light antique walnut," which I think is just the walnut color. (When there was only one color, instead of three, it was possible for the catalog to get more descriptive.)
• Shiny Walnut (Tenite), Durant Plastics #4202, 1970-1973. The walnut rearing stallion in the Durant catalog looks like a brighter and shinier tan than the typical, 1960s walnut models.

Outside the horse line, the woodcut Mustang was painted gold for beer signs in the late 1960s. (See the Chapter on "Extra Horses.") Those models were molded in white Tenite. A collector, Sande Schneider, found one that I think was painted gold (or copper brown), but was stripped down to the white plastic by a previous owner. In addition, a "driftwood" Mustang is illustrated with the "Details and Oddities."

Mold 20. The Tennessee Walker 9" series Woodcut

The Tennessee Walker engraved woodcut, which shows carving marks but not a wood grain, was made only in the 1960s. It actually measures 8" high. Sculptor Roger Williams accurately depicted the running walk gait, which is characteristic of Tennessee Walkers.

The 9" Tennessee Walker came in three colors and two mane variations. The colors are cherry, walnut, and ebony, with the amount of dark stain on the walnut Walkers varying from a lot to hardly any. On the earlier and more common type, the mane is on the left side of the neck; on the later type, the mane is shorter, on both sides of the neck, and sweeps backward. The pattern of carving marks on the body is different between the two mane types, and I've seen two different whittle patterns among the short-maned Walkers. The Tennessee Walkers are often marked with the Hartland name, but my walnut Walker with a short mane just has the letter "K."

The Walker debuted with the long mane in March 1964, and the same photo appears in three, undated, consumer catalogs from the 1964-1967 period. The shorter mane appeared in the 1965-1968 dealer catalogs, but the long-maned Walkers are more common! In 1967 and 1968 only one color was offered: "light antique walnut," which I think was just a fancier way of saying "walnut," and not a different color. Of 45 woodcut Tennessee Walkers reported in my 1995 *Hartland Market* "census," 27 were long-maned and 18 were short-maned; 21 were walnut, 14 were cherry, and 10 were ebony. There were 11 long-maned walnuts, 10 short-maned walnuts, 10 long-maned cherries, 4 short-maned cherries, 6 long-maned ebonies,

and 4 short-maned ebonies. The amount of stain on the walnut and cherry Walkers and highlighting on the ebony Walkers varies, and some have no stain or highlighting at all. Some with stain have shinier plastic than others, too. Since none are shown in the Durant catalog, this is apparently just variation within the 1960s woodcut colors by Hartland Plastics.

This model's variations are:
• Cherry, long mane on left (Tenite), # 887C, 1964-1965; dark brown.
• Cherry, shorter mane on both sides (Tenite), #887C, 1965-1966; dark brown.
• Ebony, long mane on left (Tenite), #887E, 1964-1965; black.
• Ebony, shorter mane on both sides (Tenite), #887E, 1965-1966; black.
• Walnut, long mane on left (Tenite), #887W, 1964-1965; tan.
• Walnut, shorter mane on both sides (Tenite), #887W, 1965-1968; tan.

Mold 21. The Three-Gaited Saddlebred 9" series Woodcut

Illustrating the elegant trot of the three-gaited Saddlebred, this 8.5" engraved woodcut model was made in cherry, walnut, and ebony matte finishes by Hartland Plastics and in "light antique walnut" by Durant Plastics. The 1967 and 1968 Hartland catalogs offered the Three-Gaiter only in "light antique walnut," which I think was the same color as walnut. (When only one color was offered, there was room to use a fancier color description.) The 9" Saddlebred woodcut was the last 9" series horse mold to join Hartland's stable. They do not have a Hartland mold-mark, but my matte-finish ones have the letter "Y" and my shiny-finish walnut (from Durant), has a "D" on an inside leg. Colors are:
• Cherry (Tenite), #888C, 1965-1966; appears dark brown.
• Ebony (Tenite), #888E, 1965-1966; black with subtle grayish highlights.
• Walnut (Tenite), #888W, 1965-1968; tan finish with dark stain visible in crevices.
• Shiny Walnut (Tenite), Durant Plastics #4202, 1970-1973; is a shinier and brighter tan plastic than the typical, Hartland Plastics walnuts, has very little stain, and may have rough seams or even extra plastic extending from a hoof.

The Three-Gaited Saddlebred was a very specialized model, but very appreciated by gaited-horse lovers.

Elegant Breeds. With elegant breeds in a wide range of colors, the 9" series has captured the imagination of horse lovers of all ages for 40 years.

The 11" series was the last to join Hartland's stable of 1960s horses. There were only three molds in the series then, an Arabian Stallion, Quarter Horse, and Five-Gaited Saddlebred, each measuring, at most, 10" high. Some of Hartland's most beautiful and imaginative colors are found in this series, including pearl white, glossy red bay, and blue appaloosa. If the 9" series was noteworthy for its slenderness, the 11" series was its robust counterpart.

The 11" series debuted in 1967 and Hartland Plastics made them only through 1968. Although most 1960s Hartland horse brochures were in black-and-white, it's appropriate—and fortunate—that the 1967 and 1968 dealer catalogs were in full color. The 11" series horses were known as the Regal series and Superb series. The Regals sold for $4 in those days. The Superbs cost $5 because more work went into them. Both Regals and Superbs were Tenite.

The second Hartland company, Durant Plastics, made the three, 11" series shapes in one color each from 1970-1973. Additional colors were added in the 1980s and 1990s by Steven Mfg. and by Paola Groeber. When Steven began making horses in 1983, it began with the 11" series. This chapter also includes the 11" series Grazing (Stock Horse) Mare, a new mold Steven added to the Hartland line in the mid-1980s. The Jewel and Jade molds, however, are in chapter 14.

Roger Williams was the sculptor of the three, original 11" series horses, and sculptor Alvar Bäckstrand readied the models for mold-making. Except for the Arabian Stallion, the other three molds are anatomically mares although one could argue that mares are seldom so muscular as the 11" Quarter Horse or Saddlebred, and that Hartland wasn't necessarily trying to be specific about gender. The 11" series Grazing Mare was sculpted in Hong Kong and purchased by Steven Mfg.

The horse molds detailed in this chapter, and their heights during the different production eras are:

11" series Model Heights	1960s	1970s	1980s-1990s
22. Arabian	9.5"	9.5"	9.75"
23. Saddlebred	9.5"	9.5"	10"
24. Quarter Horse (Head Tucked)	7.75"	7.75"	8"
25. Grazing (Stock Horse) Mare	—	—	6.5" at the highest point, the withers

Eye Whites. Among the 1967-1968 11" series horses, some have eye whites and others do not; the 1970s Durant models do not; the 1983-1986 styrene models do not; 1988-1990 Tenite models do not, but models by Paola in the same colors sometimes did; and the 1993-1994 Steven horses did have eye whites painted in.

Mold Marks and Boxes. In the 1960s and 1970s, the 11" series had no mold mark to identify it as a Hartland product. There are no single, molded letters on the inside leg, either. The 1980s and 1990s, 11" series models are mold-marked, "Hartland/Steven." The 11" series was Tenite in the 1960s and 1970s, styrene from 1983-1986, and Tenite again from 1988-1994. By coincidence, the Tenite models by Hartland, Durant, and Steven all had cellophane-front boxes while the styrene models by Steven had a closed, cardboard box dated 1983 with the model's picture on it.

Original Prices: 1980s-1990s. Purchased by mail from Paola, the 1980s, 11" Steven horses in styrene were $7.55 or $7.95 in 1985, depending on the color; $7.95 or $8.55 in 1986; $8.05 or $8.65 in 1987 and reduced to $6.05 or $6.65 in 1988. (Some other distributors charged about $9 for them in 1984.) Models in the higher price tier were the 11" series Grazing Mare, albino Quarter Horse, and pearl blue Arabian.

In 1988-89, Paola sold the 11" models that she produced (in Tenite) for $13.50 ($14.50 for appaloosas), and special runs were $13.95. The Steven, 1988-1990 Tenite models were about $12-$15 in stores.

In 1990, Cascade Models was selling Paola's horses by mail for $16.95, $17.95, or $18.45, depending on the color. After Paola stopped taking orders at the end of 1990, Cascade Models sold the final stock on hand in 1991 for about double those amounts.

In 1993-1994, the 11" series from Steven wholesaled for $13 and retailed for $22.50 from Cascade Models, except the Arabian Stallion and Grazing Arabian Mare were $21. In the 1994-1995 Enchanted Doll House catalogs from Manchester, Connecticut, the 11" Quarter Horse, "Prairie Dancer," was $26.

Colored Plastic. Most 11" series horses have white plastic under their paint, but some, 1960s bay Quarter Horses and Pearl White Arabians were painted over golden yellow Tenite, and the 1993-1994 #202 Dapple Rose Grey Saddlebred is painted over cream-colored Tenite; at least, mine is. The Steven Dark Bay Grazing Mare, #206, was molded in brown styrene. Other examples are noted in the lists of models. The 11" series was never sold unpainted except by Paola Groeber.

Unpainted, White Horses. Paola sold four, unpainted, white 11" series horses. They are listed here, instead of with their respective molds:

11" Arabian Stallion, white styrene, #204U; Christmas 1986, $5.50; reduced in a 1988 catalog (as #BOD1) to $2 and in another 1988 catalog, to $1.

11" Quarter Horse, white Tenite, #BOD3, 1988-1989, $5; in September 1990, $10.50.

11" Arabian Stallion, white Tenite, #BOD5, 1988-1989, $5; in September 1990, $8.

11" 5-Gaited Saddlebred, white Tenite, #BOD7, 1988-1989, $5.

11" Grazing (Stock Horse) Mare, white Tenite, #BOD11, $5; in 1989 (as #BOD9), $5; in September 1990 (as #BOD12), $8.

Unpainted models are illustrated in the Hartland Horses: Details & Oddities photo section, instead of with the photos of their respective molds. Besides the white models Paola sold, some washed up after the 1993 Steven flood. A photo in the July 13, 1993, *Hermann Advertiser-Courier* pictured Mitzi Scott, wife of Steven co-owner (at the time) Sam Scott, retrieving models from the flood water. The horse in her hand is an 11" white, unpainted Arabian Stallion.

In Hartland, Wisconsin, collector Jaci Bowman found an 11" Saddlebred in unpainted white with "AC CPL 11/6/67" written on it in red magic marker. The 1960s Hartland officials I've asked aren't sure of what the letters stood for. An 11" Arab painted silver with masked-off points remaining the white color of the acetate plastic (but hooves painted black) came up for sale in 1999. The seller said it came from a relative of a former Hartland Plastics employee. A Hartland official said there would have been no factory purpose for a model painted that way, but any color could have been used to "test" the paint mask. The model is not really unfinished because silver was not a base color for any 1960s horses. It was probably just an experiment or something an employee deliberately painted to take home with them.

Saddlery. Western saddles and accessories scaled to the 11" series were available from Paola from 1987-1990. The saddles, priced at about $40 ($50 in 1989-1990), were by Kathleen Bond of K.B. Leather Art. They came in oak leaf and traditional, floral patterns and were offered with square skirts or Arab skirts. (Arab skirts are rounded and look better on Arabs and other short-backed horses than square skirts do.) Paola sold her red roan 11" Quarter Horses only with a set of saddlery; the price was $129.

Matching breast collars, saddlebags, tapederos (ornate, stirrup covers), etc., were extra. Prices in 1990 were: parade eagle taps, $6.95; saddlebags, $11.95; regular breast collar, $12.95; thin Arab breast collar, $10.95; billets and back cinch, $8.95; and bridle with bent-wire bit, $14.95. In addition, western show halters and English/Arab show halters were $6.95. The latter came in blue, silver, red, purple, or green.

Item Numbers. In the 1960s, the 11" series horses had four- or five-digit numbers that started with "9." In 1982, I gained access to a 1968 dealer catalog that had specific model numbers and color names written in by hand for nine of the 11" series models that were identified in the printed catalog only by group numbers, and with no color names. The catalog had belonged to a former Hartland employee. It's fortunate that the information was preserved, and it appears in the model listings.

The 1970s, Durant models had numbers in the 4000s. In the 1980s-1990s, the Steven horses had numbers in the 200s while models painted by Paola had numbers in the 400s. On Paola's sales lists, note the model number to determine which models she was selling were by her and which were by Steven.

Lists of the 11" series Arabian stallion, Saddlebred, Quarter Horse, and Grazing Mare models follow.

Mold 22. The 11" Series Arabian

A marvel of muscle and spirited demeanor, this standing Arabian Stallion, robust even by 1960s breed standards, was marketed in some 1980s catalogs as a Morgan. He measures 9.5" (1960s-1970s) to 9.75" (1980s-1990s) high.

1960s, 11" series Arabian Stallions, by Hartland Plastics:
• White with gray mane and tail (Tenite), 1967, #9912, a Regal series model; white body (the color of the plastic) with gray mane, tail, and hooves. The catalog called it "white with black points," or "off-white with black mane and tail," but this model is apparently always white with gray mane and tail, gray hooves, and white lower legs.
• Black with silver pearl points (Tenite), 1967, #9913, a Regal series model; black body with silver m-t-ll-h; glossy.
• Palomino (Tenite), 1967, #9914, a Regal series model; dark yellow body with white mane, tail, lower legs, and hooves; both glossy and matte versions are found.
• Pearl White (Tenite), 1967, #9915, a Superb series model; the catalog called the color, "genuine pearl white finish"; the model is white all over except that the eyes, nostrils, and inner ears are black; apparently, the pearl white color was painted over white plastic in many instances, while others that were painted pearl white over golden-yellow plastic have a more golden look.
• Glossy, Red Bay: Blood Bay (Tenite), 1967, #9916, a Superb series model; body red as a cherry; black mane, tail, hooves, and body shadings; four white stockings; no white on face; very glossy finish; molded in white plastic.
• Sorrel: no color name in catalog (Tenite), 1968, part of the Regal series group #90021; warm brown body with beige mane, tail, and socks; black hooves; flaxen sorrel would be another name for this color. (Note: on real horses, socks are white, not beige.)
• Charcoal Gray with black points (Tenite), 1968, #99121: part of the Regal series group #90021; solid gray body with black m-t-ll-h; matte finish. In the 1968 dealer catalog from a Hartland employee, the #90021 Sorrel Arab was crossed out, and "Charcoal/99121" was handwritten. The Charcoal model did not appear in the catalog. The Charcoal was evidently a later model, and a replacement for Sorrel.
• Matte, Red Bay (Tenite), 1968, "#9913 Blood Bay" in Regal series group #90021; red body with black mane, tail, knees, hocks, and body shadings; white face and four stockings; matte finish; molded in white plastic.
• Sooty, Tan Bay (Tenite), 1968, "#99141: Buckskin" (hand written) in Regal series group #90021; body painted tan (orange brown) with black shadings, black m-t-ll-h; the painted bay/brown body color is similar to the Budweiser Brown 9" series Thoroughbred, not similar to buckskin yellow; molded in white plastic.

1970s, 11" Arabian Stallion (by Durant Plastics):
• Tan Bay without shadings (Tenite), 1970-1973, Durant Plastics #4500, no color name; orange-tan body with few or no shadings; mane, tail, lower legs, and hooves are black; black extends high, well onto knees and hocks; molded in white plastic.

1980s-1990s, 11" Arabian Stallions by Steven Mfg.:
• Black: Raven Black (styrene), Steven #204, 1983-1986; black with a blaze starting where the forelock leaves off; hooves are grayish.
• Chestnut with darker brown points: Light Bay (styrene) Steven #205, 1983-1986; reddish brown with darker brown mane, tail, lower legs, and hooves. There is an unconfirmed report that later versions of this model had black points, which would make their color bay, instead of chestnut.
• Blue with white points: Pearl Blue (styrene), Steven #209, 1985-1986; metallic blue-gray with white mane, tail, lower legs, and hooves.
• Flea Bit Grey (Tenite), Steven #204, "Royal Elite," 1993-1994; black and gray shadings over white body; tiny black flecks; mane and tail are gray; black hooves, knees, hocks, and muzzle; eye whites.

11" Arabian Stallion by both Steven and Paola:
• Rose Grey (Tenite), Steven series #262 and Paola #390, both 1988-1990; body pinkish-maroon with maroon mane and tail; black hooves, four white stockings; models by Steven and Paola look alike; Paola made about 250; those made by Steven boost the total.

11" Arabian Stallions by Paola Groeber:
• Dark Dapple Grey (styrene), Paola #223B, May 1987 test run of 70 made; body uniformly light-to-medium gray with well distributed dapples. This model can be distinguished from the Flea Bit Grey by its styrene plastic, distinctly black mane and tail, and white dapples, rather than dark specks. The Flea Bit Grey is Tenite and has a gray-shaded mane and tail.

• Light Bay (styrene) Paola #223A, May 1987 test run of 50 made; reddish brown body with black mane, tail, and hooves; black on muzzle and some black on lower legs.

Fewer than 10:
Dark Bay—styrene—red-brown body with black m-t and no white—c.1987—sold 1994—only one, Paola called him "Dandy."
Pearl White—Tenite; one sold in 1999, dated 6/1989; it's inscribed, "Sample 2."

Mold 23. The 11" series Five-Gaited Saddlebred

Hartland's powerfully muscled, 11" series Five-Gaited Saddlebred is racking. Its pose resembles that of a horse on page 220 of the 1953 edition of *The Horse*, by D.J. Kays. The photo's caption reads, "Sir James, five-gaited gelding...demonstrates his ability at the rack." Colors are:

1960s, 11" series Five-Gaiters by Hartland Plastics:
• White (Tenite), 1967, #9932, Regal series; catalogs called it, "white with black shaded mane, tail, etc.;" glossy, white body with gray mane, tail, and hooves; some have a black m-t-h, instead, but the lower legs are always white.
• Bay with black legs (Tenite), 1967, #9933, Regal series; "medium bay with black points"; body color is reddish-brown to maroon; it is not a bright red; m-t-ll-h are black; there is no white visible anywhere although the underlying plastic is white.
• Copper Sorrel (Tenite), 1967, #9935, Regal series; catalogs called it, "high sorrel with white points;" body is metallic copper color with shadings; m-t-ll-h are white.
• Cherry Red Bay (Tenite), 1967, #9936, Superb series; catalogs called it "mahogany chestnut with white socks," but it is bay because the mane and tail are black. The horse in the dealer catalog has a rich brown body color (and black m-t-h), but I think the catalog, with its odd, mustard-colored background, did not print true to color, and the bright, cherry red bay models that exist are #9936. The cherry red bays usually have black body shading that slightly mutes their bright color; m-t-h are black; the face and stockings are white. An example without shadings is illustrated, also.
• Bay with white legs (Tenite), 1967, #9937, Superb series, catalogs called it, "light chestnut with white socks" or by no color name, but it is bay because the mane and tail are black. It's body is more brown than red; it has four white stockings, black hooves, and black m-t. It is a far more realistic horse color than the Cherry Red Bay.
• Dark Bay with white legs (Tenite), 1968, "#99331: Dark Bay" (hand written) in Regal series group #90041; this model's body is more dark brown than red brown (and not anywhere near bright red); mane, tail and hooves are black; it has four white stockings and a white (bald) face.
• Dusty Bay with white legs (Tenite), 1968, "#99321: Doeskin" (hand written) in Regal series group #90041; this model's body color is like the Budweiser Brown 9" Thoroughbred color: tan paint over white plastic and overlaid with faint, black body shadings; mane, tail, and hooves are black; there are four white stockings.
• Buckskin (Tenite), 1967 or 1968; body is honey yellow muted by pale brown shadings, painted over white plastic; black points (m-t-ll-h); cannot assign a number; this model was not shown or mentioned by name in the catalogs; it was probably a mid-year substitute, but it seems to be as plentiful as the catalog colors.

11" Five-Gaiter by both Hartland Plastics and Durant Plastics:
• Orange Bay (Tenite), 1968, Hartland Plastics "#99351: Sorrel" (hand written) in Regal series group #90041; also, Durant Plastics #4502, 1970-1973; orange body painted over white plastic; usually some black shadings on knees and hocks; black mane, tail, and hooves; four white stockings. If the boundaries with black are not tidy but spill over (check the eyes and hooves, especially) the model was probably made by Durant because Hartland Plastics used masks to neatly define the painted areas; otherwise, the orange bays by Hartland Plastics and Durant Plastics can look alike.

1980s-1990s, 11" Five-Gaiters by Steven Mfg.:
• Brown: Light Bay (styrene), Steven #202 LB, 1983-1986; brown body with darker brown mane, tail, hooves, and pasterns; blaze on face; no white on legs; it's a chestnut color, rather than a bay.
• Black with white socks (styrene) Steven #230, 1983-1986; black with three white socks (all but the left hind) and blaze: the three legs with white socks also have white hooves.
• Palomino (styrene), Steven #208, 1985-1986; bright yellow body with white mane, tail, stockings, and hooves.
• Black with black legs (Tenite), Steven #201-05, a spring 1993 special for Cascade Models; 150-200 made; painted blue-black with no white;

paint may flake; no white on face. There was also a spin-off: Cascade Models painted a blaze on 33 of them for Shay Goosens' fourth annual Break A Leg model horse show in 1994. Fewer than 10 were sold at the show, one was auctioned in 1999, and the remaining pieces were offered at the 6th annual Break A Leg Live Show in Manassas, Virginia, April 1, 2000.

• Black Pinto (Tenite), Steven special run of 20 for the 1993 Model Horse Collector's Jamboree in California; their color was about equal parts white and black; each model had an intentionally different, neatly-masked marking pattern.

• Dapple Rose Grey (Tenite), Steven #202, "Mr. Majestic," 1993-1994; this model's color defies conventional horse colors: it has a cream body shaded with black and rust, especially on the shoulders and hips; dapples are found in shaded areas; the mane, tail, legs, and much of the head are black. It would be a pale buckskin (or dun), but the dark body shadings and black head don't fit; the black head is a feature of blue roan horses; a rose grey would not have a black head; and a grey or roan horse would not have cream color in its body. At least, this model isn't boring.

11" Five-Gaiter by both Steven and Paola:

• Charcoal Gray (Tenite), Steven series #262 and Paola #395, 1988-1990; body is painted charcoal gray over white plastic; black mane, tail, and hooves; four white socks, glossy finish; models by Steven and Paola look alike; about 250 were made by Paola; the ones by Steven boost the total.

11" Five-Gaiters by Paola Groeber:

• Sorrel: Light Chestnut (styrene), Paola #220B, May 1987 test color, 31 made; body painted translucent brown (dark caramel color) over white plastic; mane, tail, socks, and hooves are white.

• Liver Chestnut (styrene), Paola #220A, May 1987 test color, 24 made; body painted translucent brown; black hooves; mane and tail are brown; may have black on muzzle; four white socks.

Fewer than 10:

Pearl White with pink nose and hooves—only one—signed & dated 2/1989—sold in 1999.

Buckskin with dorsal stripe—four socks—pink left hind hoof—and eye whites—only one—a 1989 sample sold 12/1994.

Mold 24. The 11" series Quarter Horse and Appaloosa

The 11" series Quarter Horse, with its chin tucked and neck arched down, is trotting although interpretations of backing or cantering are reasonable. Collector Sandy Tomezik praised this mold as "a picture of controlled power" in her article, "Hartlands: A Fantasy in Plastic," in Shari Struzan's *The Hobby Horse*, October 1980. The mold is 7.75" (1960s-1970s) to 8" (1980s-1990s) high.

11" series Quarter Horses by Hartland Plastics, 1960s:

• 1960s White-Grey (Tenite), 1967, #9922, Regal series; catalogs called it "white with black mane, tail, etc." or "white with black points," but this model is typically white with gray mane, tail, and hooves; lower legs are white; the white color is molded in; no shadings on body; the examples with black m-t-h are less common.

• Bay with black legs (Tenite), 1967, #9923, Regal series; catalogs correctly called it "bay with black points"; red-brown to maroon body with black m-t-ll-h; it was molded in white plastic. A variation on this color (or the #9924 bay) is a model owned by Sandy Tomezik: it is bay with no shadings, painted over golden-yellow plastic, has black lower legs, and has no white anywhere.

• Superb Bay with white socks (Tenite) 1967, #9924, Superb series "light chestnut"; reddish-brown body; black mane, tail, and hooves; four white socks and white face; this model has a richer color than the 1968 Regal Bay, but is, otherwise, very similar.

• Red Bay Appaloosa (Tenite) 1967, #9926, "blood bay appaloosa," Superb series; bright red body with black shadings; black mane, tail, and hooves; black spots on a white hip blanket; some examples have white socks above black pasterns and fetlocks, which is not found in nature; in the dealer catalog with mustard-colored background, this model appears rich red-brown, but I think the catalog colors printed wrong because the models are cherry/candy apple red. A collector who is observant about color, Eleanor Harvey, said she used to own a bright, "candy apple red" bay Quarter Horse with white socks—and no appaloosa spots. It was probably a model that was intended to be the red appaloosa, but was painted differently.

• Metallic Blue (Tenite) with white points, 1967, #9927: the catalog reads: "blue roan—silver blue with white mane, tail, and stockings"; model is painted medium, metallic blue with white m-t-ll-h; one reported example was molded in pink plastic; others may be molded in white or other colors.

• Regal Bay with white socks (Tenite), 1968, "#99231: Light Chestnut" (hand written) in Regal series group #90031; reddish-brown body with black mane, tail, and hooves; four white socks and white face; this model is very similar to the 1967 Superb Bay, but its color is a less rich red-brown.

• Blue Appaloosa (Tenite), 1968, "#99271: Appaloosa" (hand written) in Regal series group #90031; dark blue body with black spots on a white hip blanket; black m-t-ll-h.

11" Quarter Horses by both Hartland Plastics and Durant Plastics:

11" Quarter Horses are found in two shades of buckskin: honey and golden tan. (Both are "flat" colors with no body shadings, dapples, etc.) Buckskin Quarter Horses in the lighter color appear in both the Hartland and Durant catalogs, but the darker color never appeared in a catalog—assuming that the catalog colors printed true. I think the lighter color (honey buckskin) was the 1960s color by Hartland Plastics, and the darker color (golden tan buckskin) was the color painted by Durant Plastics in the 1970s. I think Durant's catalogs used Hartland's photo (with the background cut out). Durant may have even sold some of Hartland's honey buckskin 11" Quarter Horses in the 1970 Durant package because Durant was, at first, selling leftover stock purchased from Hartland Plastics when it got out of the toy business in 1969. The rule of neatly sprayed eyes belonging to the 1960s models is confounded by the fact that I have one of each shade of buckskin, and the painting work is about equally tidy on both of them: each has some over spray. The models are:

• Honey (Lighter) Buckskin (Tenite), 1968, "#99221: Buckskin" (hand written) in Regal series group #90031; translucent honey yellow color painted over white plastic; the points (m-t-ll) and hooves are black.

• Golden Tan (Darker) Buckskin (Tenite), produced during about the second half of 1970-1973, Durant #4501: no color name; fairly opaque golden tan color over white plastic; the points (m-t-ll) and hooves are black.

1980s-1990s 11" Quarter Horses by Steven Mfg:

• Light Bay (styrene), Steven #200, 1983-1986; the body is painted clay-color with black points or with black mane and tail, but dark brown lower legs; some are matte where others are glossy; on all versions, the body color is even and clear with no shadings and no dapples.

• Brown Appaloosa: Bay Appaloosa (styrene), Steven #201, 1983-1986; there are two versions of this model: some are matte, orange-tan with dark brown points (m-t-ll) and hooves; others are glossy, reddish brown with darker brown mane and tail; both versions have a white blanket over the barrel and hips with spots of the body color on the blanket.

• White: Albino (styrene), Steven #207, 1985-1986; very pale, blue-white body with pale pink nose, belly, and hooves; some (later ones) have plain white, rather than blue-white bodies, and have black hooves; eyes are brown or red. Among several white or near-white models of this mold, this is the only one in styrene plastic.

• White-Grey for Jamboree (Tenite), a Steven special of 25 made for the 1992 West Coast Model Horse Collectors Jamboree; they arrived late and 20 were actually raffled in November 1992 at the Terrace model horse show in southern California, instead of at the Jamboree (the raffle included participation by mail); five were kept for later fund-raising; white body with gray shadings; no dapples; pink feet.

• Dapple Sooty Buckskin (Tenite), Steven #200, 1993-1994; golden brown body with black shadings on shoulders and hips; black mane, tail, muzzle, and lower legs except for socks in front; dapples are found in the black-shaded areas.

• Medium Dapple Grey (Tenite), Steven #200-20, "Forest Dew," 1994 special of 200 for Modell Pferde Versand, Germany, sold out by July 1995; of three 11" Quarter Horse models in dapple grey, this one is intermediate in color. It has black on the muzzle, but the whole head is not dark; the barrel is lighter than the shoulders and hips; the mane, tail, and lower legs are black except for a left hind sock. This model could be purchased from Cascade Models for $36.

11" Quarter Horse by both Steven and Paola:

• Palomino (Tenite), Steven series #262 and Paola #402, 1988-1990; golden-yellow body with white mane and tail; white stockings, black hooves, and black on muzzle; has a blaze; some were matte, but others were glossy; the body color also varied slightly: some were lighter or darker than others; about 350 were made by Paola; the ones made by Steven boost the total.

11" Quarter Horses by Paola Groeber:

• Black Appaloosa (Tenite), Paola #400, 1987-1990; black body with black spots on white, hip blanket; white face and hind stockings, and beige hind hooves; many have a pink muzzle; about 800 were made. The 1987 catalog showed a model with fewer and larger spots, but subsequent catalogs depicted a model with smaller and more numerous spots, and that type is more common. In addition, this model has mottling—small white spots near the boundaries of the white blanket, and at the muzzle and flanks; some have more mottling than others.

• Red Roan (Tenite), Paola #401, a 1988 special run of 50, sold out in 1989; body is pale maroon with dark red mane, tail, and lower legs (except for white stockings); sold with a hand-tooled, leather saddle with light walnut stain for $129; of the 50, one made for collector Barri Mayse had white sprayed on its hindquarters to make it an appaloosa; all have glossy finish; most have two hind white stockings.

• Red Roan Appaloosa: Leopard Appaloosa (Tenite), Paola #403, 1988-1990, regular run, about 400 made; body mostly white on the barrel and hips, but pinkish in other areas; mane and tail are brownish-red; brownish-red spots are scattered over the hips, shoulders, and neck; most were glossy, but one sold by Paola in 1999 was matte.

• Creamy Dun (Tenite), Paola #404, 1989-1990; cream body with black mane, tail, hooves, muzzle, and lower legs except for a left, hind sock; black shading over neck and withers and black dorsal stripe. Paola Groeber said this was her most popular model; she produced about 1,350 of them.

• Light Dapple Grey (Tenite), Paola 1989 show special, a total of about 88; a white mane and tail distinguishes this model from other dapple grey 11" series Quarter Horses; it's body is also a lighter gray, with darker gray color only on the muzzle, knees, and hocks; there are four white socks, a pink nose, and one pink hoof (the other three hooves are black). This model was made for two different model shows: 50 with eye whites, signed and dated, "Hartland Collectables, Inc., Paola Groeber 6/89," for Marney Walerius' model show in Barrington, Illinois, August 1989; and about 38 for Heather Wells' model show in Riverside, California. The model sold for $20 at both shows.

• Dark Dapple Grey (Tenite), Paola 1989 special of 50 for Marney Walerius' Model Horse Congress, Barrington, Illinois; the body is dark grey with small, white dapples distributed evenly on it; much of the head is dark, and the body color is uniform from neck to tail; it has a black mane, tail, and lower legs, except for hind socks; one hoof is flesh colored; there are eye whites painted in, and a small, narrow blaze; it sold for $20.

• Grulla, Pearled (Tenite), Paola #400P, a 1989 special run of about 90; a pearled color, this model's body is medium gray with metallic flecks in the paint; there are black points except for a sock on the right foreleg; four black hooves and black shading on the muzzle and along the crest of the neck; except for that shading, the color is even, and there is no dappling.

• Sorrel, Pearled (Tenite), Paola #400S, a 1989 special of about 40; another gorgeous, pearled color, this model's body is redder than, but similar to, the color of copper, and its paint contains metallic flecks; its mane, tail, and all four lower legs are beige although a real horse would have white stockings, instead.

• Red Dun (Tenite), Paola 1990 special, "Dun in Dreams," for Black Horse Ranch (250 made, numbered, sold for $21.75); body is glossy, pale orange; points are dark red except for a hind sock; model resembles a BHR filly born in 1990; both model and mare have a dorsal stripe, blaze, snip, and right hind sock.

• Pearl White (Tenite), Paola special of 12 painted in 1991; most were sold in 12/1994; resembles the 1960s pearl white body color; in addition, there are eye whites, pink behind the hocks and on the belly, random pink hooves (such as four pink or two black and two pink), and some have shading on the top area of the tail; they are signed and dated.

• Silver (Tenite), Paola special of 12 made about 1990 and sold 12/1994; plain silver body with black mane, tail, and hooves; this model has no shadings, no dapples, and it is not gray; numbered between 1 and 30 because 30 were made, but 18 were later repainted to other colors, leaving 12 in the silver color. (See below.)

Fewer than 10 (all Tenite):

The first six models below were created in 1989/1990 by painting a color on top of a silver undercoat; they are numbered between 1 and 30, and most were sold in 12/1994:

Sooty Buckskin with dorsal stripe and four socks—seven made.

Very Dark Dun with dorsal stripe and shoulder and leg stripes and four pearled white socks—number 1—one made.

Red Bay with two hind socks (the sales list called it "dark bay" but it is cherry red)—five made, signed and dated 1989.

Dark Brown Bay with one hind sock—number 3—one made.

Midnight blue with hind sock—three made.

Blue-Black with one hind sock—one made.

Additional models (not with silver undercoat):

Black Appaloosa with no spots on white hip blanket—two made in late 1980s—one sold 12/1994.

Black Appaloosa with 1960s Hartland appaloosa mask spots plus Paola's spots—one made in 1986 and sold 12/1994.

Pearled Grulla with light lower legs—a test color for grulla, which normally had dark lower legs; Sandy Tomezik owns it.

Red Roan with matte finish (normally, they were glossy)—two were sold in 1999, one with a chocolate saddle with oak leaf pattern, and another with no tack.

Dark Brown with white points (called "Charcoal")—three made in 1990—two sold in 1999-2000.

Metallic Gold Bay—black mane and tail—two or three samples made in the late 1980s.

Metallic Gold with white (unpainted) mane, tail, and stockings—one sample—signed and dated 4/90.

Dapple Blue—one was given as a prize at a show in St. Louis in 1990.

Metallic Copper—three made; one was raffled at Daphne Macpherson's Hartland model horse show in Washington state, November 1990.

An unusual palomino with brown hooves and no blaze (originally painted for Marney Walerius); it has slight rust-brown shadings. Marney's name for the model was "Dublin Gold." There was also a very pale (and unshaded) palomino that was a test color; there may be more than one of those.

Mold 25. The 11" series Grazing Mare

The 11" series Grazing Mare mold originated in the 1980s. It was sculpted for Steven Mfg. in China, but molded and painted in Missouri. I don't think Steven realized the horse is, except for its head and neck position, a copy of the Breyer Yearling stock horse mold. The Grazing Mare is 6.5" high at the top of the withers. Colors are:

• Dark Bay (styrene), Steven #206, 1985-1986; chocolate brown body with black points and hooves, black muzzle, and (white) blaze on face. One catalog said "light chestnut" in error. It is molded in brown plastic.

• Black Appaloosa (Tenite), 1988-1989, Steven series #262 and Paola #385; black body with white, spotted hip blanket. Both companies called the model, "Grazing POA"; POA stands for Pony of the Americas, a breed that is essentially an appaloosa pony. Models by Steven and Paola look almost alike. One of Paola's price lists read, "bay appaloosa," but it was a typographical error.

• Bright Buckskin (Tenite), Steven #206, "Gold Duster," 1993-1994; apricot orange-yellow body with black m-t, black on the hind legs and black on the forelegs above white socks and pale hooves; bald (white) face except for black muzzle.

Dramatic Models. The original 11" series is very dramatic, but collectors should keep in mind the difference in quality between styrene models and Tenite models, and not pay Tenite prices for styrene.

In the 1960s, four additional 11" series horses were planned, but never produced. They got only so far as the metal model stage, and their molds were never finished before model production ended at Hartland Plastics in 1969. Their photos are included in the Hartland History: People, Places, & Models gallery due to the thoughtfulness of collector Pam Young. The photos were taken in 1987 by collector Marney Walerius, who hosted the largest, annual model horse show in the country for two decades. The models are: a standing Quarter Horse, Thoroughbred, and Three-Gaited Saddlebred, and a Tennessee Walker at the running walk.

The popular, Lady Jewel and Jade molds joined the Hartland parade of breeds in 1988. They were the first completely new horse sculptures created for Hartland since the 1960s. (The 11" series grazing mare came first, but was not entirely original.) Lady Jewel and Jade were injection-molded in Tenite plastic. The resin horses painted and sold by Paola under the Hartland name in the late 1980s and early 1990s gave collectors some new choices and ushered in model horsedom's "resin revolution."

Models in this chapter have been made in the 1980s-1990s only:

Model Shape Name	Material	Height	Sculptor
26. Lady Jewel (Mare)	plastic	8.25"	Kathleen Moody
27. Jade (Foal)	plastic	6.5"	Kathleen Moody
R1. Miniature Horse	resin	4.75"	Kathleen Moody
R2. Peruvian Paso	resin	9.5"	Kathleen Moody
R3. Quarter Horse	resin	5"	Carol Gasper
R4. Tennessee Walker	resin	5"	Carol Gasper
R5. Arabian Gelding	resin	8.75"	Linda Lima
R6. Arabian Stallion	resin	9.25"	Kathleen Moody

Lady Jewel and Jade

Mold 26. Lady Jewel—11" series Cantering Arabian Mare

Artist Kathleen Moody sculpted Lady Jewel, a cantering Arabian Mare with a lovely, arched neck and stylized, windblown mane and tail. Lady Jewel was commissioned in the 1980s by Steven Mfg. and Paola Groeber and debuted in 1988. In the 1990s it was produced by Steven Mfg. It's mold-stamped, "Moody 88" on one leg, and "Hartland/Steven" on the other, and is 8.25" high.

Mold 27. Jade—11" series Arabian Foal

Jade is the frisky, cavorting foal created in 1988 to accompany Lady Jewel. Jade is 6.5" high, was sculpted by Kathleen Moody, and manufactured by Steven Mfg. and Paola Groeber. It is marked "Moody 88" inside the left leg and "Hartland" with the Steven logo, inside the right hind leg.

Sets of Lady Jewel and Jade from Paola:

In 1988, Paola charged $13.50 for the rose grey Lade Jewel and $8 for the chestnut Jade, but their prices went up to $16.95 and $9.49 in 1989. Since they could be purchased separately, they were not a set in the literal sense, but in every other way. Colors are:

• Lady Jewel in Rose Grey (Tenite), Paola #700, 1988-1990; body is maroon muted to grayish pink with maroon mane and tail, hind white socks with beige hooves, black hooves in front, and a white face stripe.

• Jade in Chestnut (Tenite) Paola #710, 1988-1990; this foal went with the rose grey Lady Jewel; the foal is more reddish than brown, and has four white socks and a joined star and stripe on the face; some examples have less red and more brown in their body color; mine is hand-dated 1/90 and is glossy, muted red.

• Black, Lady Jewel and Jade set with silver mane, tail, and hooves (Tenite), a June 1989 Paola special run available by mail and awarded to collector class winners at some model horse shows in 1989-1990; has blue undercoat; silver m-t-h. Collector Tina English-Wendt said she suggested this color to Paola to have a match for the 1960s, 11" Arabian Stallion in black with silver points. Some are dated "6/89" in silver on the belly. A later batch of this set dates to fall of 1991. A news note by Tina English in her *High Stepper's Review*, Sept.-Oct. 1991, said that in September 1991, Paola was "busy getting Arab resins out" and that the black Lady Jewel and Jade models were "still in the works." Shipments would not begin on those until the resins were "all [sent] out."

Rare Lady Jewel and Jade models by Paola (fewer than 10):

Transparent plastic—four Lady Jewel models (but no Jade models) were molded in transparent, butyrate plastic (like Tenite acetate) in 1987; some were sold in 1994 and 1999.

Pearled White, Tenite, with pink nose & hooves—1990—six sets.

A buckskin overo pinto set in Tenite owned by Sandy Tomezik.

Two sets made in Dark Sorrel: Charcoal in 1989: the foals were molded in cream-colored Tenite with bald—cream—face and cream m-t and four socks; the mares have m-t painted cream and two white front stockings.

Sets of Lady Jewel and Jade from Steven Mfg.

• Black Blanket Appaloosa (Tenite), Steven, no #, 1992 special of 333 sets for Black Horse Ranch; set sold for $30; they are mostly black with black spots on white hip blanket.

• Chestnut Pinto—mostly white (Tenite), Steven, no #, 1992 special of 333 sets for Black Horse Ranch; set sold for $30; they are mostly white with a few large, rounded, brown areas hand painted on the barrel, neck, etc.; they were painted without masks, so each set was somewhat unique in the placement and size of the brown markings. Rather than a common, overo or tobiano pattern, these horses are more nearly "medicine hat" pintos.

• Bay (Tenite), Steven special for 1992 JCPenney Christmas catalog. The mares were molded in brown acetate, points were painted black, and black body shadings were added. There were two foal variations: (1) molded in brown plastic like the mares; and (2) some that collector April Powell described as, "painted white [over brown plastic], then painted brown, so it appears chalky." In either case, the mane and tail were then painted black; both types are about equally common, and the bay JCPenney sets as a whole are common. They were $24.99.

• Dapple Grey Mare and Black Foal—"Bedouin Princess and Blessing," (Tenite), Steven set #233, 1993-1994. The mare's body is mostly pale gray with fine dapples in gray-shaded areas; the color is set off by black stockings, muzzle, and black on parts of the mane and tail. The foal is black with four white socks and pink hooves and a white "asterisk" mark on forehead. They wholesaled for $20 and retailed for $40 direct from Steven Mfg.; Cascade Models sold them for $35.

• Liver Chestnut Pinto—mostly brown (Tenite), Steven set #233-20, "Gloria and Nimbus," a 1994 special of 300 sets for Modell Pferde Versand, Klettgau, Germany; all suffered paint damage and models sold were recalled by MPV, but some collectors kept theirs; the mare and foal are mostly dark brown with large, white pinto spots.

• Bright Sorrel (Tenite), Steven set #941, "Selvyn [the mare] and Soliel" [the foal], a 1994 special of 300 sets for Sandy Kirch/The Equine Center, New York state; many, but not all, suffered paint damage; model is burnt-orange (warm brown) with golden-yellow mane and tail; where the burnt orange paint came off, it revealed that the golden yellow color was also the undercoat.

Fewer than 10: Steven made one, Tenite "Buckamino," Lady Jewel ("palomino" with black legs) and donated it to the 1993 Jamboree auction; Gail Berg owns it.

Unpainted, White Sets. The Lady Jewel and Jade sets were never officially sold unpainted, either by Paola or Steven, but I know of three collectors who have them in unpainted, white Tenite. The set illustrated in Chapter 8, the Hartland Horses: Details & Oddities, was purchased from Paola. Other collectors may have gotten them from Steven by special request.

Limited Edition Resin Horses

In the late 1980s and early 1990s, Paola Groeber commissioned, produced, and sold resin horses by three sculptors: Kathleen Moody, Linda Lima, and Carol Gasper. Paola had the resin models cast by DaBar Enterprises, then painted them and sold them under the Hartland name. Most of the resin models were limited editions of 50-75. They are almost unbreakable.

Mold R1. American Miniature Horse

The Miniature Horse is trotting and 4.75" high. Kathleen Moody was the sculptor, and the models are mold-marked, "KM88" on the inside of the right hind leg. In 1995, Paola said that, at most, about 150 (three runs of 50) were made. The Miniature Horse came in two production colors:

• Rich Bay (resin), "Heritage Hopeful," Paola #800, 1989.

• White (resin), "Heritage Mist," Paola #801, 1990; has a tawny mane and tail.

Fewer than 10: In spring 2000, Paola sold three samples painted about 1989: a dapple grey and two black pintos.

Mold R2. Peruvian Paso — Original name: "Oro Del Sol"

The Peruvian Paso in action, a Kathleen Moody sculpture about 9.5" high, came in two production colors:
* Sorrel (resin) "Elite El Paso," Paola #810, 1990; 35 were made.
* Dark Bay (resin), "Elite El Rio," Paola, no #; Paola painted 13 this color in November 1992 and sold them in December 1994, four years after her company, Hartland Collectables, had formally closed; model has brown body with black points (m-t-ll) and tri-color eyes.

Fewer than 10: Blue Roan—one painted by Paola as an auction item for Heather Wells' model show in California, c.1990.

Mold R3. The Quarter Horse in the Pocket Ponies series

Standing with the wind in its mane and tail, this sculpture by Carol Gasper is 5" high. It came in one color:
* Dapple Grey (resin) "Grey to Stay," Paola #850, 1990, about 50-75 made.

Mold R4. The Tennessee Walker in the Pocket Ponies series

This show-type Tennessee Walker is doing the "big lick" running walk. The sculpture by Carol Gasper is 5" high, and was made in one production color.
* Red Roan (resin), "Trip Ticker," Paola #860, 1990, about 50-75 made.

Fewer than 10: In 2000, Paola sold two samples painted in 1990: a blue-black and a dapple grey.

Mold R5. Arabian Gelding

This accurately walking, Arabian Gelding by sculptor Linda Lima is about 8.75" high.
* Dapple Grey (resin), "Shariff," Paola #890, 1990, 38 were made.
* Red Bay (resin), "Gypsy Prince," issued directly by artist Linda Lima, beginning in December 1991, for $230; 17 were painted red bay by Maggie Keene.
* "Arabian Gelding"; 28 more were painted in various colors by Linda Lima and sold by her for $230.

Mold R6. Arabian Stallion

After 12 out of 50 Arabian Geldings broke in transit from the resin caster to Paola, the Arabian Gelding was withdrawn from the Hartland line. Kathleen Moody offered to replace it with a new Arabian model, and "Simply Splendid" was born in February 1991. This cantering stallion is 9.25" high and scaled to the 11" series. All together, about 100 were produced.
* Dappled Grey (resin), "Simply Splendid," Paola, 1991; Paola painted and sold 12 in dapple grey to substitute for 12 of the Arabian Geldings; three of them were double-dappled.
* Various colors, issued directly by Kathleen Moody, with the molding, packing, and shipping done through DaBar Enterprises and Joan Berkwitz. Kathleen said that Joan Berkwitz painted about 20 in chestnut and bay; other colors were sold also.

Fewer than 10 (painted in 1990 and sold by Paola in 12/1994 or spring 2000): Bright Chestnut with four stockings—one made; Sorrel with flaxen mane and tail and four stockings—one; Pearled White—two; Blue-Black—one.

Resin Prices. In 1990, the resin-cast horses were priced as follows. The first number was Paola's price; the number in parentheses was Cascade Models' price: Miniature Horse in either color — $50 ($65);"Pocket Ponies" Quarter Horse and Tennessee Walker — $45 ($65); Peruvian Paso and Arabian Gelding — $150 ($180); Arabian Stallion, Simply Splendid — $150.

Note: After Paola Groeber's Hartland Collectables, Inc., closed at the end of 1990, the above three resin artists continued to issue new resin horses on their own. In addition, Kathleen Moody has also sculptured horses issued in plastic and porcelain by Reeves International, the parent company of Breyer Animal Creations.

Horses That Never Were. In the 1990s, Steven Mfg. intended to add seven, new equine models (to be mass produced in plastic) to the Hartland line, but fate intervened. Two of the new sculptures—a saddle Mule and a Friesian by Kathleen Moody—appeared in the 1994 Steven catalog, but financial hardships caused by the flood of summer 1993 curtailed the plans, and the models were never produced. Their molds were not even finished: The two models pictured in the catalog were wax versions of the original clay models. The other five models, created in clay only, were a Percheron by Linda Lima, and four other horses by Kathleen Moody: a leaping Lipizzan (doing the capriole), Trakehner doing the piaffe (trotting in place), walking Morab, and fox-trotting Missouri Fox Trotter. All seven clay originals were unharmed by the flood of 1993, but in fall of 1995, Steven staff discovered that they had disappeared. The models, which had been purchased outright, were not returned to the artists, and their whereabouts is unknown.

Since 1997, Breyer Animal Creations has issued a plastic Fox Trotter sculpted by Kathleen Moody. It's pose is different than the one she crafted for Hartland (Steven), but collectors can be pleased that this later sculpture is, according to Kathleen, more accurate than the one Hartland would have made.

Hartland's horse breed models actually began with horse families, which came in two general sizes: 5" and 7". In 1961, the Hartland catalog was mainly populated by horse-and-rider sets, but that's where the 5" and 7" Arabian families made their debut. This chapter covers 32 equine shapes in all: 28 horses in the 7" series and 5" series horse families, the 6" bachelor Arab Stallion, the Farm Horse and Donkey, and the Nativity Donkey. Although these models span four decades (five if you count the Nativity Donkey) and involve four Hartland companies, about half of them were made only in the 1960s. However, the quantities made in the 1960s were usually so large that many of these models are quite common. Anyone setting out to collect them will find it easy.

Hartland's horse families were sold as three-piece or two-piece sets. They include models in painted colors, woodcut finishes, and unpainted models molded in various colors. In all, there are 17 shapes of adults and 11 foal shapes in the horse families. (These totals refer to shapes only; in addition, most families were issued in multiple colors.) The breeds of family sets are:

7" series, three-piece sets: Arabian, Thoroughbred, Quarter Horse/ Appaloosa, Morgan/Pinto, Tennessee Walker, and (woodcut) Saddlebred.

7" series, two-piece sets (mare and foal only): (smooth) Saddlebred, and (woodcut) Thoroughbred.

In the 5" series, all sets were two-piece (mare and foal only): Arabian, Thoroughbred, and Quarter Horse.

The molds of horses in this chapter and their heights (in inches) and eras made are:

7" Family Series (and 6" size):	1960s	1970s	1980s-1990s
28. Arabian Stallion	6.75"	—	6.75"
29. Arabian Mare	6.25"	—	6.25"
30. Arabian Foal	4.75"	—	4.75"
31. Thoroughbred Stallion	6.25"	—	—
32. Thoroughbred Mare (grazing)	4.75"	—	—
33. Thoroughbred Foal (nursing)	3.5"	—	— (height at tail tip)
34. Quarter Horse/Appaloosa Stallion	6.25"	6.25"	6.25" test colors only
35. Quarter Horse/Appaloosa Mare	5.5"	5.5"	5.5" test colors only
36. Quarter Horse/Appaloosa Foal	4.5"	4.5"	4.5" test colors only
37. Morgan/Pinto Stallion	6.5"	6.5"	6.5" test colors only
38. Morgan/Pinto Mare	6.25"	6.25"	6.25" test colors only
39. Morgan/Pinto Foal	3.75"	3.75"	3.75" test colors only
40. Tennessee Walker Stallion	6.5"	—	6.5"
41. Tennessee Walker Mare	5.75"	—	5.75"
42. Tennessee Walker Foal	4.75"	—	4.75"
43. Saddlebred Stallion—woodcut	6.5"	—	—
44. Saddlebred Mare—woodcut	6.25"	—	—
45. Saddlebred Foal—woodcut	4.75"	—	—
46. Saddlebred Mare—smooth	6.5"	—	—
47. Saddlebred Foal—smooth	5"	—	—
48. Thoroughbred Mare—woodcut	6.25"	—	—
49. Thoroughbred Foal—woodcut	4.5"	—	—
50. Arabian Stallion—woodcut (6" size)	6"	—	—

5" Family Series:			
51. Arabian Mare	4.5"	4.5"	—
52. Arabian Foal	3.5"	3.5"	—
53. Thoroughbred Mare (grazing)	3.75"	3.75"	—
54. Thoroughbred Foal (nursing)	2.75"	2.75"	— (height at tail top)
55. Quarter Horse Mare	4.25"	4.25"	—
56. Quarter Horse Foal	3"	3"	—

Farm Series & Nativity Set:			
57. Farm Horse (some are 4.25")	4.5"	4.5"	4.5"
58. Farm Donkey	3.75"	3.75"	3.75"
59. Nativity Donkey (1950s — 4.5")	—	—	—

Years of Debut. The first horse families were the 5" and 7" Arabian families, which started in 1961. In 1962 the 7" series Quarter Horse and Thoroughbred families began and were joined by the 5" series Quarter Horse and Thoroughbred families in later 1962 or 1963. The 7" Appaloosa, Mor-

gan, and Pinto families started in 1963. The last of the horse families were added in 1965: the 7" Tennessee Walkers, woodcut and smooth Saddlebreds, woodcut Thoroughbreds, and the 6" Arabian Stallion. The horse molds that debuted in 1965 were not marked with the Hartland name.

Varied Use Histories. The molds in this chapter vary widely in their history of use, with some made in all eras, or only in the first, first and second, or first and third eras. The farm horse and donkey were made in all eras. The woodcut models and 7" Thoroughbred family were made only in the 1960s. The 5" families and 7" Quarter Horse and Morgan molds were used only in the 1960s and 1970s, but Paola issued test runs of the 7" Quarter Horses and Morgans. The 7" Arabian and Tennessee Walker families were produced in the 1960s and in the 1980s-1990s. The Nativity Donkey was made only in the 1950s, which technically expands the time range of this chapter to five decades.

Heights Remained Constant. In the 9" and 11" series, the 1980s-1990s horses are up to one-quarter inch taller than their 1960s-1970s counterparts, but for the equines in this chapter, the heights for newer and older models are about the same, or the 1980s-1990s models are ever-so-slightly smaller. Heights among 1960s horses are usually consistent, but I have a Morgan stallion in light gray styrene that is 6.75" high, instead of the usual 6.5" for Morgan Stallions. The farm horse has two sizes differing by about one-eighth of an inch, but I don't know whether the taller ones are older or newer. For the two sizes of nursing, Thoroughbred foals, the measurements are taken at the highest part of their tail.

More Styrene Than Tenite. Of the 1960s horses, woodcut models were Tenite (acetate). Among painted families, some 7" Morgan and Pinto families are Tenite, but most 7" horses and, apparently, all 5" horses were styrene, the more light-weight plastic. The 1970s family horses were styrene. The regular-run family horses of the 1980s-1990s were Tenite; the test models were styrene or Tenite.

Woodcut Horses. The six, woodcut shapes in this chapter look whittled although their original models were clay. The Saddlebred and Thoroughbred families also have horizontal streaks of engraved, dots or dashes to represent grain lines, so the catalogs always called them "woodgrains." The 6" Arabian is whittled, but doesn't have the grain streaks. The catalog sometimes called it a "woodcut" and, other times, called it a "woodgrain." I prefer to use the term "woodcut" for all six horses because it conveys the idea of a carved texture.

Painted Details. Eyes and nostrils were painted black or gray on all 5-7" series horses except woodcuts, unpainted horses, and two 7" Arabian Stallions in the 1968 catalog. Painted horses in the 1960s often had shading on each side of the face, but the 1970s horses don't. The 1980s-1990s horses usually have dark shading on the entire muzzle. Remarkably, the 1993-1994 Steven Arabian and Tennessee Walker families have eye whites painted in. It's the first time models smaller than the 9" series were ever given eye whites.

Unpainted Models. In the 1960s and 1970s, unpainted models in the 7" size and smaller were normal, catalog items, a finished product with smooth seams. A few were sometimes sold in plastic bags or on a blister card, but they were usually sold individually and unpackaged, in a cardboard bin on the counter at the dime store. (I bought some that way, myself.) The unpainted models from the farm size through the 7" family size were always styrene, and they are common. Unpainted models were molded in many colors of plastic, not just white.

Unpainted horses made by Hartland Plastics in the 1960s and Durant in the 1970s can look almost alike, but the 1970s models are more likely to have slightly mismatched seams or glue visible where it spilled over. The glue was clear, but turned yellow over time.

In the lists of models, I usually designate the unpainted models simply as "a budget item," because the 1965-1969 catalogs had 11 overlapping groups of unpainted animals, and the models in each group cannot always be identified. Many were sold loose in bins, and are sometimes pictured in a jumble, with their heads or feet sticking out of the bin. Contents of the groups could vary from one catalog to another, and identifiable horses are often found in several different groups, all with cumbersome names such as "#1002—Animal Friends Counter Display." In addition, Durant's 1970-1973 catalogs had three, unpainted animal groups. While I give the dates for unpainted horses as 1965-1969, some were still being sold in 1970-1973 and additional models in some of the colors may have actually been produced (by Durant) during 1970-1973.

It's a tribute to the quality of the sculpture that the family series horses even—or especially—look good unpainted.

Painting Methods. Two methods were used to achieve the colors on painted horses. Some models were entirely painted while others made use of the molded-in color of the plastic and were painted only on the extremities and detail areas (mane, tail, lower legs, eyes, shading on face, etc.).

Boxes. In the 1960s, the 7" horse families came in boxes with scenic backdrops. The Quarter Horses had a ranch scene with a brown fence and cattle. The backdrop for the Appaloosas and Pintos was a mountain range. The Thoroughbreds, Morgans, and Tennessee Walkers had a farm scene with white fences; the woodcut Saddlebreds, a sketch of horse show action; and the Arabians, a desert vista.

The boxes could be hung on the wall. Instructions on the back read:

Hang from here after punching a small hole in front cover. This carton makes an ideal shadow box display for your horse collection. Simply turn over the front cover and slip this section into it. Can be hung on the wall or set up on shelf. Use small tacks for wall display and hang from punched holes.

The 5" series horses were usually sold on blister cards if painted, and loose in a bin if unpainted.

Original Prices. In the 1960s, the three-piece families (except woodcuts) were $2.98, the woodcut Saddlebred families were $3.98, the two-piece Saddlebred set was $1.98, the woodcut Thoroughbred set was $2.98, and the woodcut Arabian was $1.49. The two-piece, 5" series horse families were 98 cents. Unpainted horses cost less: I paid 19 cents for a 5" series Arab Foal in golden yellow in 1967; the 7" series adults were 49 to 59 cents during 1965-1973.

For her 7" families, Paola charged $19.50 in 1988 and $24.50 in 1990 ($25.90 for the Pearled White Tennessee Walker Family). In 1993-1994, the 7" Arabian Family wholesaled for $18; it retailed for $30 from Cascade Models. The two-piece, palomino Tennessee Walker mare and foal set from the 1993 JCPenney Christmas catalog was $29.99.

Catalog Item Numbers. In the 1960s, the horse families were numbered in the 600s or 6000s. In the 1970s, Durant numbered them in the 4000s. Paola's 1986-1990 7" series models were numbered in the 500s, except for her 1986 test runs numbered in the 200s. Steven's 7" series horses were numbered in the 200s.

The list of models begins with the 7" Arabians.

Molds 28-29-30. The 7" series Arabian Family

Stallion—6.75" high, Mare—6.25", Foal—4.75"

The three-piece, 7" Arabian Family consisted of a stallion standing elegantly with its left hind leg bent at the fetlock and a standing mare and foal, all with serene expressions. The 7" Arabians began in 1961. Hartland Plastics always numbered them as set #676 regardless of color, but the names given to the horses did change. In a 1961 catalog, the 7" Arabians were named, "Silk, Satin & Velvet," but a January, 1961 price list named them as, "Sheik, Sheba, and Shah." Those names are found on some boxes of the gray Arabians (bodies painted gray with black points), but other gray sets came in a box naming them, "Sultan, Sama, and Sultana," and those names also appear on the box of gray Arabians illustrated in the February 1962 *Western Horsemen* magazine.

The 7" Arabian family pictured in the 1961, Hartland dealer catalog in full color was a tri-color family (named "Sultan, Sama, and Sultana") with: Buckskin Stallion (yellow body with black points, and the eyes are painted black); Mare in gray-blue with white mane and tail and black lower legs; and Foal in white with black points. No one has ever reported a tri-color set, but entire sets in blue and in white with black points are found. I only know of one collector with a buckskin Arabian set (with the golden color painted over white plastic). The stallion and foal look authentic, but the mare might be customized. Arabian stallions in buckskin with the golden, body color molded-in (and oddly enough, eyes not painted) were in the 1968 catalog. The 7" Arabians were also sold unpainted in golden-yellow plastic, which some kids or adults could have customized into "buckskin." Factory models were neatly airbrushed.

Unpainted 7" Arabians were made in several colors in the 1960s. The Arabian Family does not appear in the 1970-1973 Durant Plastics dealer catalogs, but Paola and Steven made beautifully painted, 7" Arabian families in the 1980s-1990s.

Colors of 1960s, 7" Arabian Families:

• Gray (glossy), 1961 or 1962-1964: Arabian stallion, mare, and foal with bodies painted gray over white plastic and points painted black; most of the painted 7" Arabians from the 1960s are this color. The gray Arab set appeared in the 1962 and 1964 Red Bird sales Co. catalogs, the April 1963 Frederick C. Wolf & Son wholesale catalog, and in the 1963 and 1964

Insemikit Co., Inc. (of Baraboo, Wis.) catalogs. I saw the gray Arabs at Woolworth's. The retail price was $2.98.

• White with black points: molded in white or ivory plastic with neatly airbrushed, black m-t-ll-h and painted eyes; there are no gray body shadings except a little on the face; this color was probably a variation on the usual, 1962-1964, gray Arabian family with black points. It is less common than the gray-bodied ones.

• Blue (gray-blue): a 1960s color not in catalogs, but sets in the box have turned up (collector Peggy Howard has one); Arabian stallion and mare with blue-gray body, white mane and tail, and black lower legs; the foal is blue-gray with black points.

• Bay with white stockings: seen in the 1963 and 1964 Insemikit Co. catalogs for $2.98; Arabian stallion, mare, and foal with maroon-brown body with black m-t-ll; black hooves, blaze on face; glossy.

• Bay with black points; maroon-brown paint over golden yellow plastic with points painted black; a stallion that looks authentic is illustrated in this book, and a mare and foal are shown in a 1960s-era Hartland brochure promoting its molding services; the color is not common.

• #6761, White-Grey—Stallion only—white with gray mane and tail and black lower legs; the mane and tail paint areas were airbrushed, but not masked; eyes are not painted, and that is how the model is shown (with unpainted eyes) in the spring 1968 catalog; was sold on a card for $1.00 as #6761. No matching mare and foal were made.

• #6761, Buckskin—Stallion only—model is molded in golden yellow plastic with mane and tail painted gray (airbrushed, but not masked), lower legs painted black, and the eyes not painted; spring 1968, sold on a card for $1.00. No matching mare and foal were made.

• Buckskin with body color painted-on over white plastic: 1960s, scarce at best.

• White or ivory (unpainted); styrene, 1965-1969.

• Pale Taupe/dark oatmeal (unpainted); styrene, 1965-1969.

• Golden Yellow (unpainted); styrene, 1965-1969.

1980s-1990s, Regular Run 7" Arabian Families:

• Rose Grey (Tenite), Paola # 510, 1987-1990; three-piece set with grayish pink (muted maroon) bodies and mane and tail a little darker.

• Light Dapple Grey (Tenite) Paola #511, 1988-1990; was called "dark dapple grey" in the 1988 catalog, and the earlier ones were reportedly a little darker although the same photo was used in each catalog.

• Liver Chestnut (Tenite) Steven #271, Tenite, Steven, 1993-1994; dark brown with golden beige mane and tail, four white socks, and pink feet; foal has a white "asterisk" on its forehead.

Test Run, 1980s-1990s 7" Arabian Families:

• Light Bay (styrene), Paola #213HB (same # as Light Grey), Christmas 1986, advertised as a test color of 21 sets; light, reddish brown with black mane, tail, hooves, low black socks, and black on the knees and hocks.

• Light Grey (not pearled, styrene), Paola #213HB, Christmas 1986, 13 sets. (This model has the same number as the Light Bay, due probably to a mix-up on the order form. Note that in Paola Groeber's article in the July/August 1987 *Model Horse Gazette*, she corrected the quantities of models sold at Christmas 1986. Not all models were painted at the time the sales list was mailed, and she painted the balance of the models to accommodate the orders. Thus, the quantities sold at Christmas 1986 usually differed from the quantities advertised. The Light Bay Arabian sets were omitted from the corrected list, so an update on their quantity was not available.)

• Light Grey (Pearled, styrene) Paola #213HP, "Pearl Grey," Christmas 1986, 14 sets; the pearled finish gives the dapples a "light sheen," Paola wrote.

• Dark Grey (Tenite) Paola #213TB (same # as Raven Black), Christmas 1986; 19 sets; it was originally called "light grey" on the order form; these models have (tiny) dapples.

• Raven Black (Tenite) Paola #213TB (same # as the Dark Grey), Christmas 1986, 16 sets; no white on all or most of these models.

• Sorrel: Light Chestnut (Tenite), Paola #213TC, Christmas 1986 test color; the color is brown with white mane and tail and some white socks; the order form stated that there was one set, but the corrected number was 11 sets, with variable white socks.

Fewer than 10:

Chestnut or Dark Rose Grey (Tenite); rich, rosy brown with mane and tail matching the body. About 1987, Paola made "a handful" of sets in dark rose grey before changing to the lighter, rose grey color illustrated in the catalogs.

Paola painted speckles on a 7" foal molded in black butyrate (Tenite acetate-like) plastic in 1985 and sold it in 1999.

Steven test-molded a few 7" Arabian families in black Tenite in the 1980s—their seams were not cleaned, so they still look rough. Sandy Tomezik owns a set.

Molds 31-32-33. The 7" series Thoroughbred Family

Stallion—6.25", Mare (grazing)—4.75", and Foal (nursing)—3.5" at highest part of the tail

The three-piece, 7" series Thoroughbred family consists of a walking stallion, a grazing mare, and nursing foal. It has been made in only one painted color: the very common, glossy, maroon bay. The 7" Thoroughbred family appeared in the Sears Christmas catalog in 1962 and 1963 with white stockings, but the models are always bay with black lower legs. Their Sears price was $2.69 in 1962, but they went up 10 cents in 1963; the typical retail price was $2.98. The 7" mare and foal are similar in pose to the 5" series Thoroughbred mare-foal set, which are also very common. No 7" series Thoroughbreds have been produced since the 1960s.

Copies. Hong Kong copies of the Thoroughbred stallion are found in palomino, dark gray, and according to collector Jaci Bowman, in bay. She got her gray stallion at a Ben Franklin store. A palomino stallion owned by collector Laura Whitney is mold marked, "Hong Kong," with linked diamonds, a "P" in one diamond, and a "C" in the other.

Hartland 7" Thoroughbred Family colors are:
* Maroon Bay, Hartland Plastics #673, 1962-1964; maroon with black shadings and gloss coat; black points; glossy.
* Black (unpainted, styrene): Stallion is very dark gray; the mare and foal are black; the colors are molded in; 1965-1969.
* Golden-yellow (unpainted, styrene): I found a stallion this color in 2000.

Molds 34-35-36. The 7" Appaloosa and Quarter Horse Families

Stallion—6.25" high, Mare—5.5", and Foal 4.5"

The three-piece sets of Quarter Horse and Appaloosa families are from the same molds: a standing stallion, a mare with head bowed and right foreleg lifted, and a foal with its ears back and left foreleg raised. Whether the mare and foal are standing (restlessly) or walking is debatable, but the mare is in a nice trail class pose. The pose of the mare and foal is similar to their counterparts in the 5" series. The 7" Quarter Horse family is pictured in the 1970 Durant catalog and mentioned in the 1971-1973 Durant catalogs. In the 1980s, only test colors, no production runs, were made.

Copies. Hong Kong copies of the 7" Quarter Horse mare and foal are found in palomino, dark gray, and bay. The palomino mare and foal are mold-stamped, "Hong Kong," high on the inside right leg. Copies of the mare and foal are also found in metal in the 5" series size.

The Hartland 7" series Appaloosa and Quarter Horse families are:
* Appaloosa family, Hartland Plastics #681; 1963-1964 and 1968; dark gray with white hip blanket and black spots on the blanket and in the gray areas on the barrel, neck and shoulders; glossy, retailed for $2.98. Some have a darker gray body than others and fewer spots on the barrel.
* Buckskin Quarter Horse family, Hartland Plastics #672, 1962-1964; golden yellow-tan body with black points, molded in golden plastic; there are two variations of this color: on the more common type, the body color is simply the color of the golden plastic (and points are painted black), there are no shadings except for a little dark gray shading on the face; on the less common type, golden shadings and a gloss coat were added to the golden-plastic body (and points were painted black). They were $2.98.
* Bay/Sorrel Quarter Horse family, Hartland Plastics #900, Silver Canyon Roundup set. The SCR set was made in 1965, and still available for $4.98 in the 1966 Mission House catalog. The set includes a bay (brown body with black points and no white) 7" Quarter Horse Stallion and Foal and a sorrel 7" Quarter Horse mare (brown body with beige m-t-ll-h). Those colors appear in the 1965 Hartland dealer catalog. The 1966 Mission House catalog shows a bay mare and sorrel foal, but those colors have not been reported. Collector Sandy Tomezik has two examples of the stallion: one painted bay over golden plastic and the other painted bay over reddish brown plastic. The Silver Canyon set also included a 7" western series Wrangler horse with molded-on saddle and rider with hat and rope molded-on.
* Golden Yellow (unpainted, styrene), 1965-1969.
* Reddish Brown (unpainted, styrene), 1965-1969.
* Durant Buckskin Quarter Horse family, Durant series #4400; listed in the 1970-1973 catalogs, but only pictured in the 1970 catalog, with dark (not golden) shadings all over. In reality, the Durant Quarter Horse family has no shadings at all or just faint, dark shadings in low areas near the elbow and stifle. They are molded in golden styrene with the black points airbrushed and masked but untidy. Unlike the 1960s Quarter Horse family, there are no shadings on the side of the face although the eyes, nostrils, and inner ears may be neatly airbrushed. A model with low socks, sputtery paint, and an extra tab of plastic on the bottom of the horse's foot is by Durant. They are toys lacking collector appeal.

1980s Test Color, 7" Quarter Horses and Appaloosas
* Palomino (styrene), Paola #212HP, Christmas 1986, 12 sets with random white markings.
* Appaloosa (styrene), Paola #212HW, Christmas 1986, 20 sets; white with random black spots, gray manes and tails, and gray shadings on face, knees, and hooves; pearl paint was applied, and then a gloss coat.
* Black (Tenite) Paola #212TB, Christmas 1986; random blazes and stockings; 10 sets.
* Chestnut (Tenite), Paola #212TC, Christmas 1986, 18 sets; described by flyer as a rich chestnut with slightly darker m-t, light gray muzzle, varied blazes and stockings, and pale yellow hooves below white stockings.

Molds 37-38-39. The 7" series Morgan and Pinto families

Stallion 6.5" high, Mare 6.25", and Foal 3.75"

The Morgan and Pinto families used the same three molds: a mare and stallion standing squarely and a foal with its head bowed and left foreleg raised, perhaps pawing. In the 1960s, Hartland made only copper sorrel Morgans and black pintos, and sets of both breeds can be found in Tenite plastic although there are about four sets in styrene for every Morgan or Pinto set in Tenite. In the 1970s, Durant Plastics made the pinto family and the Morgan family, but the Morgans were bright orange-red, instead of "copper." Some Morgans in test colors were made in the 1980s, but no regular runs have been made since the 1970s. Colors are:
* Copper Sorrel Morgan family (found in both Tenite and styrene), Hartland Plastics #677, "sorrel with flaxen manes," 1963-1968; coppery, metallic orange body color with a light coat of black shadings; white points, hooves, and blaze; molded in white plastic; were $2.98.
* Black Pinto family (found in both Tenite and styrene), Hartland Plastics #680, 1964-1968; mare is white with black patches and black socks; the stallion and foal are black with white patches and white socks; the pinto pattern is about the same on all sets—it's very fanciful! The "Tobiano Pinto Family" appeared in the 1963 and 1964 Sears Christmas catalogs for $2.79 and $2.89, respectively. Normally, they retailed for $2.98.
* White or Ivory (unpainted, styrene), 1965-1969.
* Reddish Brown (unpainted, styrene), 1965-1969.
* Light Gray (unpainted, styrene), 1965-1969; an attractive, dove gray color.
* Orange-Red Sorrel Morgan family (styrene), Durant series #4404, 1970-1973; bright orange-red with white points; no shadings on body; black eye color is over sprayed, seams may be rough.
* Durant Black Pinto family (styrene), Durant series #4404, 1970-1973; less tidy paint work and seams are the sign of Durant, but it can be difficult to always tell the Durant from the 1960s Hartland.

1980s, 7" Morgan Family Test Colors:
* Mahogany Bay (not pearled, styrene) Paola #211HB, Christmas 1986, 17 sets; rich, dark brown body color with black points and facial shadings.
* Mahogany Bay (Pearled, styrene) Paola #211HP, Christmas 1986, 12 sets; color like #211HB, but a coat of pearl paint was added before the gloss coat was put on.
* Black (not pearled, Tenite) Paola #211TB, Christmas 1986, 12 sets.
* Black (Pearled, Tenite) Paola #211TP, Christmas 1986, 15 sets; painted black with pearl paint and then gloss coat added.

Fewer than 10:

Chestnut (Tenite) with blaze and four stockings—Paola Christmas 1986—pictured in flyer—one set only.

A Morgan Stallion in burnt sienna (sorrel with flaxen points)—by Paola—dated 8/20/86—owned by Tina English-Wendt.

Green—foal only—unpainted styrene sample shot—sold by Paola—12/1994.

Steven test shots in black Tenite, 1980s—a few exist; Sandy Tomezik owns a set.

Molds 40-41-42. The 7" series Tennessee Walker Family

Stallion 6.5" high, Mare 5.75", and Foal 4.75"

The mare and stallion of the 7" Tennessee Walker family illustrate the fluid, running walk gait, which has the same order and timing of footfalls as a flat walk, but has a large overstep with the hind feet.

The 7" Tennessee Walkers were not made in unpainted colors, and were not made by Durant. They were produced in the 1980s and 1990s. Colors are:
* Palomino/Bay/Sorrel Tennessee Walker family (styrene), Hartland Plastics #684, 1965-1967. The 1960s Tennessee Walker stallion is a translucent, lemon-yellow palomino, and the mare, called "dark bay" is the maroon bay color with shading. The foal is sorrel with flaxen mane, tail,

stockings, and hooves. The flaxen stockings would not be found in nature, but are pretty. The models are styrene with a glossy finish. The stallions were molded in white plastic; many mares and foal were molded in maroon or brown plastic, respectively, but other mares and foals were molded in white plastic and then painted maroon or brown. (Among the painted mares and foals, the body color on some has turned dark olive green. Collector Jo Maness reported a set that color in the early 1980s.) The Tennessee Walker family was sold in the 1966 Penney's Christmas catalog for $2.22; their price was typically $2.98. The 1960s Tennessee Walker Family was not mold-marked "Hartland," and has no letters on the inside leg.

• Blue Roan (Tenite), Paola set #500, 1987-1990; body painted medium gray with black points and mostly black head; each horse has one or two white stockings and a narrow blaze.
• Light Bay (Tenite), Paola set #501, 1988-1990; reddish brown bodies with black points except for two or three white stockings per horse; narrow blazes.
• White-Grey (Pearled, Tenite), Paola set #500P, a 1989 special run; Paola thought that only about 25 sets were made; these models are a more opaque white than the 1960s pearl white color; they have some gray on the mane and tail, black muzzles, and black hooves.
• Palomino Mare and Foal (Tenite) Steven set #269-14, at least 1,100 were made for the 1993 JCPenney Christmas catalog, selling for $29.99. They had a smooth, semi-gloss finish, but the rigid, plastic bands holding the models in the box rubbed the paint; if returned to Steven Mfg, they were repainted, and came back with a more matte and textured-looking palomino color, and were shielded from the fasteners by soft, foam wrap.
• Palomino Stallion (Tenite), Steven #269-20, 1994 special run for Black Horse Ranch, which offered them only to special customers; quantity unknown, possibly 200; this is an opaque, matte, and textured palomino color with orange on the knees and hocks.
• Dark Sorrel: Liver Chestnut (Tenite), Steven #271-20, 1994 special run for Black Horse Ranch; 200 sets were made, of which about 100 sets were sold and 100 were returned to the factory due to paint damage.

1980s-1990s, Test Color 7" Tennessee Walkers:
• Blue Roan (not pearled, styrene), Paola #210HB, Christmas 1986, 11 sets; about four sets were a darker blue roan, and seven sets were a lighter shade; all have black points and no white except for eye whites.
• Blue Roan (Pearled, styrene), Paola #210HP, Christmas 1986, 19 sets; a pearl coat preceded the gloss coat; all have black points, dark shading on the face, and no white except for eye whites.
• Blood Bay (Tenite), Paola #210TB, Christmas 1986, 24 sets; these have no white markings (but do have eye whites), and are a deeper bay than the regular-run Light Bays, #501.

Fewer than 10:

Chestnut (Tenite), Paola #210TC, Christmas 1986, three sets; rich chestnut color with slightly darker mane and tail, body shadings, random white markings, and eye whites.

Wedgewood Blue—blue with white points—1990 sample—one set sold in 1999—two to four sets or more were made altogether; signed and dated 3/1990.

Dark Gray—almost black—one mare and one stallion—sold by Paola—12/1994.

Very dark blue roan, mare only—sold by Paola—12/1994—one only.

Cremello, mare only (Tenite)—donated by Steven to 1993 Jamboree auction—one only.

Black Pinto, mare and foal only (Tenite)—donated by Steven to 1993 Jamboree auction—one mare-foal set only.

Molds 43-44-45. The Woodgrain 7" Saddlebred Family

Stallion 6.5" high, Mare 6.25", and Foal 4.75"

The 7" Three-Gaited Saddlebreds (with roached mane and shaved tail dock) came in three, simulated wood colors, and show whittle marks and grain marks. The stallion and mare are both trotting, but the stallion lifts his knee higher. The foal stands with right foreleg bent, tail raised, and head tipped up. The 7" stallion is very similar in pose to the 9" individual series trotting woodgrain Saddlebred. This family has not been made since the 1960s, when it retailed for $3.98. They are not mold-marked with the Hartland name. Colors are:
• Cherry Woodgrain (Tenite), #685C, 1965-1966; dark brown.
• Ebony Woodgrain (Tenite), #685E, 1965-1966; black.
• Walnut Woodgrain (Tenite), #685W, 1965-1967; tan.

Molds 46-47. The 7" Three-Gaited Mare and Foal—with smooth finish

Mare 6.5" and Foal 5" high

This mare and foal are the same shapes as the engraved woodgrain 7" mare and foal, but are slightly larger and have a smooth surface. The mare is maroon bay with dark shadings, and the foal is sorrel with flaxen points. This foal, the woodgrain Saddlebred foal, and the Tennessee Walker foal are Hartland's most graceful foals. A smooth stallion was never made. The molds for this mare and foal haven't been used since the 1960s. They were usually $1.98, but were $1.44 in the 1966 Penney's Christmas catalog. They are not mold-marked with the Hartland name, but my mare has an "X" on the inside of the right hind leg, and my foal has a "Z." It is possible that yours might have different letters.

• Bay Mare/Sorrel Foal (styrene), Hartland Plastics set #764, 1965-1968; called "Peacocks of the Show Ring" in the 1965 dealer catalog; the mare is the shaded, maroon bay color; the foal is sorrel with flaxen points and hooves; both have a glossy finish.

Molds 48-49. The 7" scale Woodcut Thoroughbred Mare and Foal

Mare 6.25" high, Foal 4.5"

Another single-parent family, this woodcut set includes a foal in the same shape as the Tennessee Walker foal, but with carving marks and a subtle woodgrain, instead of a smooth surface. The mare stands almost squared and, in build, reminds me of the porcelain Boehm Adios model and the plastic Breyer Adios. (Both of those models came out in 1969, after the Hartland Thoroughbred.) The hardest Hartland horse molds to find, this set has not been made since the 1960s, when it was $2.98. I wrote in my first edition (1983): "Two collectors, Sandy Tomezik and Laura (Hornick) Behning reported woodcut Thoroughbred mares lacking the Hartland seal, but they are definitely Hartlands and may be languishing unrecognized in attics across the country." Since then, Hartlands in general have been coming out of the attic in record numbers. Apparently, none of the woodcut Thoroughbreds are marked "Hartland" or have leg letters. Colors are:
• Cherry Woodcut (Tenite), #683C, 1965 only; dark brown.
• Ebony Woodcut (Tenite), #683E, 1965 only; black.
• Walnut Woodcut (Tenite), #683W, 1965 only; tan.

Mold 50. The 6" Woodcut Arabian

Stallion only, 6" high

The proudly standing 6" Arabian has carving marks, but not grain marks, so is a woodcut, but not a woodgrain (as Hartland defines them). He is the only horse in his scale, and the only Arabian in a simulated wood finish. A bachelor, a family was never created to go with him. Like all of the woodcuts and woodgrains, he's Tenite, and should have survived the years well. He was one of the last horse molds to debut, and has not been made since the 1960s. You could buy him then for only $1.49. He isn't marked with the Hartland name, but may have the letter "A" on an inside leg. Another one of mine has an "H." He was given three finishes:
• Cherry Woodcut (Tenite), #610C, 1965-1966; two types: a dark brown matte finish, generally due to black stain over maroon plastic; and a shiny maroon finish due to no stain on the maroon plastic.
• Ebony Woodcut (Tenite); #610E, 1965-1966; two types: matte black (with faint, grayish highlights in the crevices), and shiny black with no hand-rubbed highlights; both are molded in black plastic.
• Walnut Woodcut (Tenite), #610W, 1965-1966; again, two types: matte tan with dark stain rubbed into the recessed areas, and shiny tan (with no stain); both are molded in tan plastic.

Molds 51-52. The 5" series Arabian Family

Mare, 4.5" high and Foal, 3.5"

This standing mare and foal (in series #6002) was made from 1961-1973, and is very common. Hartland called the 5" Arabs "Silk & Satin." Durant called them "Silk N' Satin." They were issued in several painted and unpainted colors; be sure to distinguish between factory-painted colors and unpainted models whose child or adult owners customized them by adding paint to the mane, tail, etc. Factory models were neatly airbrushed. The 1961 catalog showed two of the 5" series Arabian Mare-Foal sets: a set in bay, and one in gray-blue with white mane and tail. The blue-gray foal's lower legs look black, but the mare's lower legs match her blue-gray body. The company probably changed its mind after the catalog was printed because no 5" Arabians in the blue-gray color have ever been reported. Factory colors are:
• Gray with black points; shown in 1962-1967 catalogs (Hartland and other); molded in white plastic with points painted black and gray shad-

ings applied to the body; their blister package is dated 1961. In the 1962-1964 Red Bird Sales Co. catalogs, the gray Arab Mare & Foal were #675G.
• White with black points; shown in 1968 dealer catalog; also shown in 1970-1973 Durant catalogs (see below).
• Bay Mare and Foal; copyright on package reads 1961; the set is common; red-brown body color is molded in; points are painted black. In the 1962 Red Bird catalog, this set was #675B.
• Bay Mare and Sorrel Foal, 1960s; this color pair has been found sealed in the package (dated 1961); there was never a sorrel mare. The sorrel foal was molded in brown plastic and its points were factory-painted white.
• Palomino; this color was mentioned, but not pictured, in the 1962 Red Bird catalog as set #675P; dark golden body color is molded in and points are factory-painted white.
• Buckskin, 1960s; molded in light golden plastic and points are painted black; has not been found in catalogs, but is reasonably common. Their package is dated 1961.
• White or ivory (unpainted), 1965-1969.
• Pale Taupe/dark oatmeal (unpainted), 1965-1969.
• Golden Yellow, 1965-1969; two shades: one lighter, like the buckskin 5" Arabs; and one darker, like the palomino 5" Arabs.
• Reddish Brown (unpainted), 1965-1969.
Durant White with black points (styrene), Durant series #4201, "Silk N' Satin, "1970-1973; the black socks are lower, there is no shading on the side of the face, and the airbrush-painting work may be less neat than on the 1960s models.

Horses for Nylint. Hartland supplied horses and cows for Nylint toy truck sets for a while in the 1960s. Hartland's national sales manager, Paul Champion, recalled flying through the fog on one of his trips to Nylint Corp. in Rockford, Illinois.
The 5" series Arab mares and foals appear in several Nylint catalogs: 1961—all-yellow mare and foal in cover photo and sets in bay and palomino in an artist's rendering; 1962—bay and palomino mares in two paintings; 1963—all-yellow and bay mare and foals in cover photo and in painting; 1966—includes a 5" mare painted as a zebra; 1968—artist's drawing is of a light bay mare and dark foal with white socks; those were not actual Hartland colors. Later on, Nylint used a mare and foal pair that are roughly similar to the Hartlands, but less graceful.
Copies. A copy of the 5" Arab foal has been found in black, white, and electric yellow by collector April Powell; it is almost the exact same size and shape as the Hartland, but the legs are thinner in some places, and the eyes slant upward. Another copy of the foal, in golden-yellow styrene, has very thick legs and is thickened up all over. Some very cute, mini-sized copies of, or take-offs on, the 5" series Arab mare are also found.

Molds 53-54. The 5" series Thoroughbred Mare and Foal

Mare (grazing), 3.75" high; Foal (nursing), 2.75" high to the highest part of the tail
The second family in the series #6002 is the Thoroughbreds. The 5" series grazing mare and nursing foal show about as much detail as their 7" series counterparts. Colors are:
• Bay (Glossy, Shaded, Maroon Bay), Hartland plastics set series #6002, "Streak N' Fleet," #675T in 1964 Red Bird catalog; 1963-1969; very common.
• Golden Yellow (unpainted), 1965-1969; common.
• Cranberry (unpainted), 1965-1969; a brown-red color with a hint of magenta; less common than golden yellow.
• Unshaded Bay, molded in maroon styrene with mane, tail, feet, and low socks airbrushed black; there are no body shadings and there is no gloss coat; often, a tab of plastic extends from the bottom of one hoof of the mare. The Durant series #4201, "Kandy N' Dandy," pictured in a package in the 1970-1973 catalogs looks like the 1960s bays, but they were probably leftover models from the 1960s. Another pair of models pictured outside the package in the Durant catalog has few or no body shadings, and the eyes are unpainted. The actual models produced by Durant during 1970-1973 are this unshaded bay, but the eyes are usually painted.

Molds 55-56. The 5" series Quarter Horse Mare and Foal

Mare, 4.25" high; and Foal 3"
The 5" series Quarter Horse Mare is posed like the 7" series Quarter Horse Mare, but the foal differs from its larger counterpart by having its head bowed down, instead of up. This set in buckskin is common; bay and unpainted colors are not common. Catalogs sometimes called the buckskin color "claybank." The 1960s, 5" series horses of all three breeds had the same number, #6002. The 5" series Quarter Horse molds haven't been used since the 1970s.

The models are:
• Buckskin, Hartland Plastics set series #6002, "Sugar N' Spice," #675Q in 1964 Red Bird catalog, 1963-1969; molded in golden yellow plastic with brown or tan body shadings and points painted black. Some of these models were still being sold (by Durant) in the early 1970s.
• Bay (Shaded, Maroon Bay), 1965, mare and foal sold on a "Sunny Acres Farms" card (all eight "Sunny Acres Farms" cards of farm people, animals, and implements were #7000/7001); maroon body color is molded in; there is no gloss coat.
• Buckskin with no body shading and low, black socks. The Durant series #4201, "Sugar N' Spice" set in the 1970-1973 catalog (in a Circle H package) looks the same as the 1960s Hartland set, with golden body shadings and black points over molded-in, golden body color. However, Durant was selling 1960s stock leftover from Hartland Plastics. The models with low, black socks and no body shading or face shading were the actual version produced during 1970-1973 after the 1960s stock sold out.
• White or ivory (unpainted), 1960s; these are not common.

Unpainted Summary. Here's a summary of which 5" and 7" series horse families were sometimes sold unpainted, as budget items (in the 1960s-1970s):
7" Arabian, Morgan/Pinto, Quarter Horse/Appaloosa, and Thoroughbred families — Yes, they were sold unpainted.
7" Tennessee Walker, woodcut Saddlebred, smooth Saddlebred, and woodcut Thoroughbred families, and 6" Arab Stallion woodcut— Not sold unpainted.
5" Arabian, Thoroughbred, and Quarter Horse families — Yes, they were sold unpainted.
Note: When the 6"-7" woodcuts had no stain, they were sold at the same price as the stained woodcuts.

Mold 57. The Farm Horse

The 4.25-4.5" high standing farm horse was always unpainted and made of styrene. They are not marked Hartland, nor do they have letters on the inside leg.
The farm horse reminds me of how Saddlebreds were sometimes portrayed (with head high, but chin tucked) in old paintings or lithographs, but many collectors consider it an Arabian. The Hartland Arabians—the Tinymite, 6", 7", 9", and 11" —all have their chins in the air, but this model doesn't, which is why I disagree. Either way, it is too elegant for the Hartland farm animal set of six animals, #2002, and the #10021 card it came on with a donkey and sheep for 79 cents. The #10020 Little Red Barn set included two farm horses. The farm horse was sold from dime store bins, in the #1010 Horse & Farm Animal Counter Display, from 1965-1968 for 29 cents.
Next on its resume, it joined Durant's #4126 Farm Animals set, 1970-1973. Then, it was in the Steven #280 12-piece Farm Animal Play set, 1985-1986. Steven also made a few test shots in black, white, and brown Tenite plastic; collector Sandy Tomezik has one of each.
A similar horse, but much less graceful, accompanied a 1961 Nylint truck set. It is styrene, not marked, and was not made by Hartland. I only mention it because some observers might mistake it for a Hartland. Colors of the Farm Horse include:
• White (unpainted)
• Cream (unpainted)
• Pale Taupe/dark oatmeal (unpainted)
• Golden Yellow (unpainted)
• Plum Brown (unpainted)
I noticed that my golden yellow and plum brown colors have shorter ears than the others. They are 4.25" high compared to 4.5" for the others.

Mold 58. The Farm Donkey

The 3.75" farm donkey followed the same path as the farm horse, being made, in styrene, on and off from 1965-1986. It was in the #2002 farm set along with the horse, a 2.75" high sheep, 3.5" dairy cow, 3.25" beef cow, and 2" pig. The #10020 Little Red Barn set included a decorated, corrugated barn with a 22" corral and two of each of the animals in the #2002 set.
The farm donkey was part of Durant's #4900 Horse and Farm Animal Counter Dump Display, in which each unpainted animal sold for 19 cents to 59 cents. It was also part of Durant's #4905 Horse and Farm Animal Counter Dump Display of unpainted animals priced from 19 to 59 cents each.

The farm Donkey is not marked Hartland and has no letters on the leg. Colors of the Farm Donkey include:

- Light Gray (unpainted)
- Golden Yellow (unpainted)
- Pale Taupe/dark oatmeal (unpainted)
- Cream (unpainted)
- White (unpainted)
- Plum Brown (unpainted)

Mold 59. The Nativity Set Donkey

This 4.5" high donkey, in Tenite, accompanied the Hartland Nativity sets in the early-to-mid 1950s. Although many nativity set members are marked with Hartland's "Diamond 'I'," the donkey isn't marked. It's typically painted dark gray; earlier ones were unpainted white. Collector Sandy Tomezik used to own one in swirled colors (chiefly gray, but including white, black, and red). One that looks molded in dark, gray-blue plastic with no paint added appeared at an Internet auction in 2000. The Nativity Donkey's relaxed stance, with ears at half mast and the right hind foot at ease, shows Roger Williams' flare for animal sculpture. It's a suitably humble beast.

Catalog Descriptions. The 1960s catalogs pointed out how the models depicted their breeds well. Paul Champion and/or Robert McGuire wrote the following:

Of the 7" Arabians: "Note the dished face, the high forehead (or Jibbah), the widely flared rib cage."

On the 7" Thoroughbreds: "Racers, hunters, jumpers all have added to the classic form of this kingly breed. See the long neck, long powerful legs, high croup."

"American Quarter Horses. Powerful—quick on feet and reflex—gentle and intelligent. Note the short neck and small head—the powerful shoulders and hind quarters. Top animals posed for these animals."

In the 1990s, sculptor Alvar Bäckstrand said that, to aid their work, he and Roger Williams visited horse farms.

Families Are Popular. The Hartland horse families are popular because of their nice poses, good breed type, interesting colors and finishes, and because they are families.

The only draft horse in the Hartland model line is found in the Tinymite series. The Tinymites were six breeds of horses under three inches high. They were made for nine years, 1965-1973: from 1965-1969 by Hartland Plastics, and from 1970-1973 by Circle H (Durant). The breeds are: the Arabian, Belgian, Thoroughbred, Quarter Horse, Tennessee Walker, and Morgan. Although they are between 2" and 2.75" high, the Tinymites show a lot of detail for their size. They were sculpted by Roger Williams and modified by sculptor Alvar Bäckstrand.

Hartland Plastics started out with six molds, then modified them so they would mold more easily. The older versions have ears pricked and separated (except for the Morgan) and good body contour. In the newer versions, the ears are joined or laid against the neck, and the bodies are flatter. (The Morgan had ears parallel and joined-looking in both the newer and older versions.) The change came relatively early, and the newer-mold type is about three times as common. The old molds began in 1965; the Arab and Thoroughbred I bought at Christmas 1966 were the newer molds. In the Arabian, there is also an intermediate version.

In the 1970s, Durant used the newer molds, too, so a total of 18 different, painted Tinymite horses were issued altogether. Hartland also made unpainted Tinymites in the molded-in colors of yellow and brown. No Tinymite horses were made in the 1980s or 1990s. The Hartland name does not appear on any Tinymites, but they have a letter or two on an inside hind leg that was used to identify molded sections to the assemblers.

The Tinymite Horses, and their Heights (rounded up or down to the nearest one-quarter inch) are:

Mold No. and Breed	Old Mold	New Mold
60. Arabian	2.5"	2.75"
61. Belgian	2"	2.25"
62. Thoroughbred	2"	2.25"
63. Quarter Horse	2.25"	2.5"
64. Tennessee Walker	2.5"	2.75"
65. Morgan (head tucked)	2"	2.25"

Painted-series Tinymites seem to always be styrene, but since they are solid plastic, not hollow, they don't sound as "tinny" when you tap them with your fingernail as styrene models usually do. Painted Colors of Tinymite Horses are:

Arabian
• Sorrel, old mold, 1965; molded in brown plastic with flaxen (beige) points; leg letter: A or F.
• Sorrel, intermediate mold, 1965; same color and shape as old mold, but the tail shape is different—noticeably more straight and rounded at the bottom—instead of curling to a point. Collector Eleanor Harvey also noted that the direction of the hair lines on the tail is different from both the old and new Arabs. Thus, this was a transitional design, and not a piece that happened to come out of the mold badly.
• Sorrel, new mold, 1966-1969; same color as old-mold sorrel; leg letter: F. Collector Marcia Minor owns all three types.
• Orange Sorrel by Durant, new mold, 1970-1973; body painted orange over white plastic; points are painted beige; leg letter: a faint F without the top.

Belgian
• Bay, old mold, 1965; molded in maroon plastic with black shadings on body and black points; leg letter: broken H.
• Bay, new mold, 1966-1969; same color as old-mold bay; leg letters: JL or G.
• Orange by Durant, new mold, 1970-1973; body painted orange over white plastic; the eyes, ears, and nostrils are not painted any color but orange; some have shadings over the orange color; leg letter: G.

Thoroughbred
• Buckskin, old mold, 1965; golden-beige body with black points; leg letter: D or E.
• Black (painted), new mold, 1966-1969; model is painted black over white plastic; on mine, the molded letter on the inner hind thigh is "I" without the crossbars.

• Black (painted) by Durant, new mold, 1970-1973; painted black over white plastic; has the letters "LJ" molded on an inner hind thigh.

Quarter Horse
• Black (painted), old mold, 1965; painted black over white plastic; leg letter: B or C.
• Buckskin, new mold, 1966-1969; golden-beige body with black points; the black stockings may be higher than on the Durant (below), and there is a slight difference in the golden-beige body color: the Hartland color is slightly more muted than the Durant; leg letter: H or K.
• Buckskin by Durant, new mold, 1970-1973; golden-beige body with black points over white plastic; some are bright yellow; leg letter H or K.
• Black (painted), new mold, by Durant, 1970-1973.

Tennessee Walker
• White with black points, old mold, 1965; molded in white plastic and points painted black; leg letter: G.
• White with black points, new mold, 1966-1969; same color as old mold; mine has the letter "D" on inner hind thigh.
• White with black points by Durant, new mold, 1970-1973; looks like the new-mold Tenn. Walker by Hartland Plastics, except that the black stockings can be lower; mine has the letter "C" on an inner, hind thigh.

Morgan
• Palomino, old mold, 1965; body painted golden yellow over white plastic; the points are white; mine has letter "J" on inside hind leg.
• Palomino, new mold, 1966-1969; same color as old-mold palomino; mine has letter "E" on inner hind leg.
• Palomino by Durant, new mold, 1970-1973; it's a more pale yellow color than the 1960s models; painted light yellow over white plastic; the points are white; they have the letter "B" on an inside hind leg.
• Black by Durant, painted over white plastic. Collector Eleanor Harvey recently found one still sealed in its 1970 Circle H box marked 29 cents. Leg letter: B.

Unpainted Tinymites. I've found new and old versions of all six breeds in both brown and yellow Tenite, except the new-mold Belgian in yellow and old-mold Belgian in brown, but they are probably out there.
Unusual Colors. Collector Jacqueline Tierney has an ivory-white, unpainted Belgian and Quarter Horse (with new-style ears) that are styrene. The Belgian has the letter "G," and the Quarter Horse has the letter "H," so these models were probably from the late 1960s. Lorraine Mavrogeorge reported an Arabian painted "dark cherry red" with cream points; it was a 1960s model since its box read, "Hartland Plastics Div. of Amerline Corp. Hartland, Wisconsin."
Copies. The Tinymites have been copied, and turn up in various colors. Some copies in plastic have no mold marks at all, while other copies read, "H. K." for Hong Kong. Copies cast in metal, some painted and others unpainted, can also be found. I've seen a metal copy of the Tinymite Morgan marked "Durham Industries, Inc. 1976/made in Hong Kong."
Colors and Mold Marks (if any) on Tinymite copies (in plastic) include:
Arabians: (1) a black Arabian with "old ears" and no letter markings is apparently a copy; it has a different tail tip and a steeper and higher arch on the top of the tail. (2) a new-mold copy painted gray over white plastic with no letter or mold mark; collector Jenny Lambert said she bought it about 1988 at a Ben Franklin store.
Belgians: (3) a new-mold copy painted dark gray over white plastic, with no letter or mold mark. (4) a new-mold copy, molded in brown with black m-t-h and no shadings and no mark of any kind.
A Thoroughbred copy (5), molded in brown with black m-t-h and no shadings and an "I" or "1" on the inside, left hind thigh. The Hartlands have a letter "I," but this model looks too rough to be the Hartland with black paint added. It is the new mold.
Morgans: (6) a copy painted electric yellow—with no letter or mold mark; from a Ben Franklin store about 1988. (7) copies molded in brown, styrene plastic with points painted black; marked with H. K. (for Hong Kong) and a "diamond-P" mark.
Nylint Horses. Tinymite horses or copies of them came with the Nylint Corp. "Pony Farm" truck set #8000 in 1965-1970 Nylint Toys catalogs. Since the catalog illustration is a painting, not a photograph, and the horse images are cut out from the background, it is difficult to be certain of

their shapes, but the horses resemble the old-mold Tinymites with separated ears. Models resembling the Arab, Morgan, and Thoroughbred are reddish brown; those resembling the Belgian, Quarter Horse, and Tennessee Walker look sorrel or dark palomino at the Nylint web site. Hartland did supply Nylint with horses for a time, but whether that included Tinymites is hard to say for sure. Nylint also purchased animals made in Hong Kong, so they could be copies.

Two Molds of Each Tinymite Horse

Why Two Versions. In 1990, Hartland sculptor Alvar Bäckstrand explained the technical reason for the change from old to new Tinymite horse molds. "The 2.5" horses were made of solid plastic if I remember right." [They weren't hollow like all the larger Hartland horses.] "The reason for the filled [joined] ears was that it took one less side slide in the mold to get them out of the mold. Those molds with side slides were a whole lot more expensive than the ones with only a straight pull. It's not possible to pull the plastic out of the mold if there is the slightest undercut." So the earlier molds, which yielded nicer horses, were more difficult to work with, and the company replaced them.

Old Molds Have Separated Ears. The old Arabian's head was turned slightly to the left, while the new mold faces straight forward and is slimmer. The old Arab's tail curled up at the end, while the new mold has a straight tail. An intermediate version has the pricked ears, but a straighter tail. Old and new Belgians were about the same except for forward, separate ears on the old ones and backward, joined ears on the new ones. The old Thoroughbred had upright, separate ears, while the new one's ears are forward and joined. The old Quarter Horse had upright and separate ears; the new one had ears turned back and joined with mane between them. The old Tennessee Walker had sideways, upright ears, but the newer ones had their ears laid back. The Morgans, with neck arched, chin tucked, and ears laid back on both versions, are the hardest to tell apart. On the old Morgan, the bottom edge of the mane weaves a little compared with the smooth edge on the new version. The lower left hind leg on the old Morgan is also more vertical, and it has finer lower legs and hooves, especially in front; on the new Morgan, the lower legs were thickened up. The old Morgan legs also look "lazier" and there is a jog in the profile of the back side of the tail. The old molds are more highly prized by collectors.

Breed Switch. While the old molds were in use, the stocky, Quarter Horse mold was painted black and identified as the Thoroughbred and the lanky, Thoroughbred mold was painted buckskin and identified as a Quarter Horse. The new-mold Thoroughbreds and Quarter Horses were correctly painted and identified. Buckskin is not a Thoroughbred color.

Mirror Larger Horses. An interesting thing about the Tinymites is that they mirror some of the larger Hartland horses. The old- and new-mold Tinymite Arab Stallions look a little like the 11" and 9" Arab Stallions, respectively. The Morgan looks like the neck-arched-down, chin-tucked prancing horse of the 9" horse/rider series and the 11" series Quarter Horse. The Tennessee Walker and Thoroughbred have counterparts in the 9" Individual series; only the Belgian and the Quarter Horse have no larger look-alikes.

The Belgian is trotting; the Arab, Quarter Horse, and Thoroughbred are more or less walking; and the Tennessee Walker is doing a running walk. The Morgan is trotting, although interpretations of backing or cantering are also fairly reasonable.

Were 29 Cents. The Tinymites were the idea of Paul Champion, Hartland's national sales manager from 1948-1966. They were a modestly-priced alternative to Hartland's large horses. Tinymites sold for 29 cents in individual packages in the 1960s; unpainted ones were sold loose, in a counter-top, cardboard bin for 10 cents or grouped in sturdy, plastic bags. The 1965 or 1966, undated Mission House catalog offered them for 29 cents each or $1.70 for all six old-mold Tinymites. The catalog number for painted-series, Hartland Tinymite horses in single packages was #6004; the number for sets of three was #6008. The Durant (Circle H) number was #4000 if packaged singly, #4050 in double-packs, and #4100 in four-packs. Durant Tinymites were 29 cent each, 59 cents for two packs, and 98 cents for four-packs. After Tinymite production ended in 1973, they lingered in stores for a while. I bought several four-packs at a neighborhood pharmacy in the summer of 1975.

Nice things can come in small packages.

The Hartland line of dogs includes 12 Tinymites, two farm set dogs, Roy Rogers' dog, Bullet, and a lying down German Shepherd. The dogs and their heights in inches (rounded up or down to the nearest one-quarter inch) are:

Large Dogs:

D1. Bullet	4"
D2. German Shepherd (Lying Down)	3.75"

Tinymite Dogs:

D3. Beagle	2"
D4. Black Labrador (Labrador Retriever)	2"
D5. Chesapeake Retriever	2"
D6. Cocker Spaniel	2"
D7. Collie	2.25"
D8. English Pointer	2.25"
D9. German Shepherd	2.5"
D10. German S. H. (Short-Hair) Pointer	2.25"
D11. Golden Retriever	2.25"
D12. Irish Setter	2.25"
D13. St. Bernard	2.25"
D14. Std. (Standard) Poodle	2.25"

Farm Dogs:

D15. Adult Farm Dog	2.25"
D16. Farm Puppy	2"

• **Bullet.** Roy Rogers' German Shepherd dog, Bullet, was designed to accompany the 9" series horse-and-rider set, but was sold separately for 49 cents for eight years from 1956-1963. He's a nicely detailed dog in a standing pose, 4.25" high and 6.25" long from nose to tip of tail. Bullet's shadings, including the dark "saddle" area on his back, vary from dark gray to black. His unpainted areas are white or off-white, the color of the acetate (Tenite) plastic. He was sold in a plastic bag or came in his own, small box and had a neck tag. His name was spelled correctly on the box, but the neck tag spelled it, "Bullett," with two t's. Bullet was not shown in Hartland western catalogs, but appeared in the 1956 Wisconsin Deluxe Company toy and novelty catalog, the 1959 Miles Kimball catalog, the 1963 Hartland "Heirloom Treasures" catalog of cake-top decorations, and in the 1963 Maid of Scandinavia spring catalog.

• **The Lying Down Dog.** In 1995, collector Sande Schneider brought to my attention a lying down dog with a hollow bottom—like some of the Hartland religious figures—and which she felt matched the artistic style of Bullet. I relayed her description—a lying down dog larger than Bullet—to Paul Champion, who was Hartland Plastics' national sales manager from 1948-1966. He said he remembered a "sitting down" dog, a shepherd or a collie that was brown and about 6" long. He said it was "hollow" before I mentioned it. He said it wasn't popular. He didn't recall the dog's name, if any. The dog wears a molded-on, studded collar, and sits like a sphinx (with its belly on the ground and forelegs extended forward). It is not illustrated in any known sales literature, but it was item #196 in a Hartland archive photo from about 1950. Another Hartland official said he believed the dog was the first of a possible series of dogs of different breeds, but it did not sell well, and the idea was abandoned.

The Tinymite Dogs

During 1966-1969, Hartland Plastics made 12 breeds of "Tinymite Barkies," 2" to 2.5" high. Paul Champion said they were his idea, and he regretted that they did not sell well. They are less common than the Tinymite Horses.

The Tinymite dogs were sold in individual, cellophane-front packages for 29 cents each as #6007, and in sets of three as #6009. Then, Circle H/Durant produced them from 1970-1973. In the Durant catalogs, they were #4001 singly, #4051 in a double-pack, and #4101 when four-to-a-package.

The colors of the 12 dogs in the Durant catalogs match the 1960s Hartland colors. However, the Durant Tinymite horse colors departed from the catalog, and the dog colors sometimes did, too.

In the 1960s, Hartland Plastics painted each of the 12 breeds of Tinymite Dogs in one color scheme. They are:

• Beagle: black shadings over orange-red color on the head, neck, and back; white everywhere else.

• Black Labrador: black with smooth coat; painted black over white plastic. The breed name is the Labrador Retriever, which comes in the canine colors of black, yellow, and chocolate; Hartland made it only in black.

• Chesapeake Retriever: painted reddish-brown with shadings; has a textured coat. The Black Labrador and the Chesapeake Retriever are similar in

pose. The breed's formal name is Chesapeake Bay Retriever since it originated on the Maryland coast, near Chesapeake Bay.

• Cocker Spaniel: painted paler yellow than the Golden Retriever.

• Collie: black and brown-orange over the head, neck, and back; white everywhere else.

• English Pointer: white with black spots.

• German Shepherd: painted black on the back and ears; pale taupe (grayish tan) elsewhere.

• German S. H. (Shorthaired) Pointer: white with red-brown spots; on some of the models, the spots are chocolate brown without the hint of maroon.

• Golden Retriever: painted golden-yellow.

• Irish Setter: painted red-brown with some dark brown shadings.

• St. Bernard: black, orange, red, and white; same color scheme as the Beagle.

• Std. (Standard) Poodle: white; eyes and nose are black. Of the four styles of poodle clips illustrated in *The Complete Dog Book* by the American Kennel Club, this model exhibits the "English saddle clip," a clip used on adult show dogs.

Known, Durant colors of Tinymite Dogs from 1970-1973, with less precise painting than the Hartland Plastics models, are:

• Beagle: orange with white underside, legs, and tail top; some black on back and head top.

• Collie: orange with white head and underside; black on top of back.

• Irish Setter: orange like the Durant Belgian horse Tinymite.

• St. Bernard: black (painted over white plastic).

In addition to the painted Tinymites, some unpainted Tinymite dogs were sold. Collector and dog fancier Rick Van Etten reported an unpainted, Tinymite German Shorthaired Pointer molded in yellow plastic. The unpainted dogs are less common than the painted-series dogs, which are molded in white styrene plastic. Like the Tinymite horses, the Barkies are not mold-marked with the Hartland name.

Of the seven classes of dogs recognized by the American Kennel Club, the Hartland Tinymites included seven breeds of sporting dogs, two herding dogs (the Collie and German Shepherd), one hound (the Beagle), one working dog (the St. Bernard), and one non-sporting breed (the Poodle). No toys or terriers were made.

The Farm Dogs

Hartland Plastics made an adult dog and puppy for its "Sunny Acres Farms" series of shrink-wrapped, carded farm toys in 1964 and 1965. The farm dogs are the same size as the Tinymites, but in action poses. Both have brown topsides and white everywhere else. Rick Van Etten identified them as generic, farm type collies, maybe part border collie. He said that such dogs are usually black-and-white, though. The dogs are:

• Adult Farm Dog: 2.25" high; was part of the #707 Farmer's Daughter set along with a girl, lamb, and kid.

• Farm Puppy: just under 2" high; was part of #708 Farmer's Son set with a boy, goat, and rake.

Like the Tinymites, the farm dogs are not marked "Hartland."

Nylint Copies. Nylint Corporation, Rockford, Illinois, sold copies of the Tinymite dogs and farm dogs with some of its truck sets. A Hartland official recalled supplying Nylint with styrene "horses, cows, and sheep," in the 1960s. A Nylint Corp. official said that in the 1960s, Nylint bought horses and cows from Hartland for a while, and later bought them from overseas. However, farm people and dogs used with Nylint farm truck sets were made in Hong Kong, he said.

Copies of the Tinymite Beagle, Collie, and German S. H. Pointer accompanied the Nylint truck set #1710, "Pet Mobile" in 1968. The copies are marked, "Hong Kong" and are one-quarter of an inch taller than the Hartlands. For example, the Hong Kong Collie is 2.5" high, compared to 2.25" high for the Tinymite Collie.

The copies of the farm dogs are marked, "Nylint Hong Kong" along the lower left side of the body. The Nylint and Hartland farm dogs are the same size and shape, but the Nylints have painted eyes and mouths, their "white" areas are ivory, and the brown marking over their back has a different outline. The Hong Kong version of the farm puppy went with the 1964-1965 Nylint "Fun on the Farm" #7100 truck set. The adult farm dog was copied, also, giving collectors more things to look for.

Companions. Compared to Hartland's Tinymite horses, the Tinymite dogs and farm dogs are made in a larger scale. The dogs are, more or less, sized to be companions for Hartland's 5" series horses.

This chapter includes three horse shapes that were neither part of the western series, nor the breed series. I call them, "molds #E1-E3." One was an early horse by Hartland. The other two were advertising items molded for clients; they are not marked with the Hartland name because they were not Hartland products as such. In addition, a few of Hartland's regular horses were used for point-of-purchase displays, too, and they are described here.

Regular Horses

• **Mold #E1. White Horse on a Base**. An early, novelty horse manufactured by Hartland Plastics is 4.75" high, molded in white plastic, and attached to a base 1.25" high and almost 6" long. The base, molded in brown plastic, has no words on it, but it is marked with the Diamond "I" symbol, which means that its mold originated earlier than 1954. The horse was #B194, and painted chestnut in a c.1950 Hartland archive photo. Without the base, it was #194. Collectors have found it only in white, though.

Edwin Hulbert, general plant manager of Hartland Plastics from 1951-1972, said he thought the white horse was molded before his arrival. Hans Seuthe, the molding foreman who began in 1949, did not recall the horse, either.

• **5" Arabian Mare (mold #51) as the Butazolidin Horse**. Hartland used its 5" Arabian Mare for a Butazolidin counter display for Jen-Sal. The horse is molded mostly in clear, styrene plastic with yellow and white filler and the name of the equine pain medication in large letters on its side. The tail and right hind leg are molded in opaque, white plastic. The "Bute" horse came on a green, plastic base. It appears in a 1960s, custom-molding brochure from Hartland Plastics.

• **5" Quarter Horse Mare (mold #55) in an Old Milwaukee Sign**. A sign for "Old Milwaukee On Tap" appears to include the 5" series Quarter Horse mare. It is plated in gold color and stands at the upper right corner in the compartment, 18.5" high x 29" wide x 4.5" deep sign. The sign does not light up. Its style is probably late 1960s. Until 1983, Old Milwaukee beer was a product of the Jos. Schlitz Brewing Co., Milwaukee, Wis. Hartland officials could not specifically recall this sign, but did not completely rule it out. Hartland probably made at least the horse.

• **Gold Mustang (mold #19) in Mustang Malt Liqueur Signs**. Hartland officials said that this sign was made in its entirety by Hartland Plastics. It used the 9" series, woodcut Mustang, painted entirely in metallic gold. The two that I have are not marked Hartland: one has an "X," and the other has an "S." The same horse was used on two versions of the sign: one is an electric, wall sign with a lighted base under the horse; the other is a counter display with the horse and a bottle attached to an unlit base. The horse has holes drilled in its right side or in its feet, depending on which sign it went with. The name, "Mustang Malt Liqueur" was first used in commerce in March 1965; the liqueur was then a product of Pittsburgh Brewing Co., Pittsburgh, Penn. Hartland began making point-of-purchase displays for clients about 1965, and it advertised custom molding services through 1976. This sign was probably made between 1965 and 1969.

• **Black Mustang (mold #19) for Superba Tie Display**. A black, woodcut 9" Mustang by Hartland is attached to the base in front of a 24" high tie display rack for Superba, 100% polyester ties. Ties can be draped over a plastic horse shoe at the top of the display. Superba neckties date back to 1902, but the polyester ones are more recent. This sign was probably made between 1965 and 1969. In November 1998, this tie display with black Mustang sold at eBay (a public auction site on the Internet) for $50.

Horses Manufactured Solely For Clients

In the 1960s, Hartland Plastics manufactured two additional horse shapes that were used only for beer signs and brewery souvenirs of Anheuser-Busch, Inc., owner of the famed Budweiser Clydesdales. The two items, a single Clydesdale and a team, were made about 1965 (between 1965 and 1970). The signs and souvenirs were available to the public, and some are still available, through the Anheuser-Busch gift catalog. Some signs old enough to vote may still be on display in taverns.

• **Mold E2.** The 7.5" high Clydesdale with **right, hind leg lifted entirely clear of the ground.**

I call it the "Two-Point" Clydesdale to distinguish it from two types of 7.5" non-Hartland Clydesdales: the Toe-Touch and Flat-Foot Clydesdales. All three have a molded-on harness, bay or brown color, and the same action pose—except for the position of the right hind leg! Hartland's Two-Point Clydesdale trots with its right, hind leg lifted almost three-quarters of an inch off the ground (base). If the right, hind foot touches the ground in any way, it is not the Hartland Clydesdale. (On all three types, the left knee is raised high, and the other two feet are flat on the ground.)

There are two mold variations in Hartland's Clydesdale, with the difference found in the molded-on harness: (a) some have flat (non-working) rein loops on the harness collar (in the shoulder area); and (b) others have raised (working) rein loops and a rein hole through the mouth. Type (a) came first.

As a subcontractor, Hartland supplied Mold E2 horses for Budweiser signs and souvenirs. They were styrene plastic, which could withstand the heat in a lit sign better than acetate. Even so, some of the Clydesdales enclosed in signs warped over many years of use. The single, Hartland Clydesdale (Mold E2) is fairly common, but the 7.5" horses it didn't make are even easier to find. (See sidebar.)

• **Mold E3.** Golden Hitch with eight Clydesdales and a wagon, 12" long overall. (The entire sign is 15" long.)

This sign with a mock, beer bottle is shown in Hartland's custom-molding brochure. It is marked, "Thomas A. Schutz Co., Inc., Morton Grove, Ill." Hartland's Edwin Hulbert explained that Hartland Plastics was the subcontractor for the entire sign. After being molded in plastic, "the metalized Budweiser teams were done in aluminum-silver and then lacquered gold," he said. The Milwaukee company that did the plating used a vacuum metalizing process.

Only this size of golden team was made by Hartland; gold or silver teams shorter or longer than this length were not by Hartland. Both Edwin Hulbert and sculptor Alvar Bäckstrand said that they worked on only one size of small hitch plus the large [7.5"] single Clydesdale. Idella Williams said that her husband, Roger Williams, sculpted a small Budweiser model, 2" high. It may have been for this hitch.

Clydesdales Are Common. Plastic Clydesdale models—whether brewery-related or not—are common. The list of Clydesdale items Hartland did **not** make includes:

1. Toe-Touch Clydesdales, which have the right, hind toe touching the ground, were brewery souvenirs. The 7.5" size is marked, "Made in Hong Kong 224," and some bear a Busch Gardens sticker. Their mold was evidently an altered version of the Two-Point mold, with the right, hind toe lowered. A 3.75" size is also found; both sizes have a short (rolled) tail, molded-on harness, and are bay or brown.

2. Flat-Foot Clydesdales, which have the right, hind foot flat on the ground, were toys or novelties, not brewery souvenirs. The 7.5" size is marked, "Goldenplum, Made in Hong Kong, 225." Others are 3.5" and 3" high; all three sizes have a molded-on harness and are bay or brown, but they have long tails and loose manes.

3. Clydesdales that match the Hartland Clydesdale, but are only 3.5" high, were not by Hartland. They have the two-point pose, molded harness, bay color, rolled tail, and mane bobs. They came in sets of eight on brewery signs of both straight and circling designs. Both the brewery and Hartland officials said that Hartland did not make them or their signs. (Somewhat similar horses, 4.5" high, came in the scale model kit, "Budweiser Clydesdale 8-Horse Hitch," from AMT/The Ertl Company, Inc., Dyersville, Iowa.)

4. Among gold or silver teams of eight Clydesdales in brewery signs, if the horses and wagon measure only 7.5" long overall (from the nose of the front horse to the back of the wagon) or measure more than 12" long overall, Hartland did not make them.

5. Silver, cantering Clydesdales 4.75" high in Bud Light bar signs were not made by Hartland.

Comparing the Details on the 7.5" High Clydesdales. The Hartland Clydesdale has a tiny, Anheuser-Busch eagle on the tip of the harness collar, but the Toe-Touch Clydesdale does, too. Some Hartland Clydesdales have a rein hole through the mouth and working (raised) rein loops on the harness collar while others do not. In contrast, all of the Toe-Touch models have them. The Flat-Foot models have the mouth hole, but not the raised loops. The Hartland and Toe-Touch models have mane bobs, but the Flat-Foot Clydesdales have a loose mane. All of the above horses are styrene; most were molded in white plastic, but some of the non-Hartlands were molded in brown or golden-yellow.

The Hartland-Produced Single Clydesdale

Hartland's Edwin Hulbert said that over 100,000 horses were produced from "that single, two-cavity mold" for the 7.5" high Clydesdale with right hind foot lifted entirely clear (Mold E2). Hartland made the first 1,000 directly for the brewery, for "Gussie Busch to award to the employees on the occasion of the ?? millionth barrel of beer!" to roll out of the St. Louis brewery, he said. (Later, Hartland was the subcontractor supplying the Clydesdale to point-of-purchase design companies.) The single Clydesdale was also given to brewery employees on retirement, and was available to the public. It sold for $5 at Busch Gardens in Tampa, Florida, but was already sold out when I checked in 1982.

The Clydesdale was used in many designs of signs and souvenirs. An early design Hartland produced is more or less a paperweight. The horse is mounted on a black, plastic base 13" x 5.25" x 1" high that reads, "Famous Budweiser Clydesdale Horse." A label on the bottom of the base (on mine) reads, "Thomas A. Schutz Co., Inc., Morton Grove, Ill." Hartland Plastics was the subcontractor. Hartland made both the horse and the base, but in all other cases, supplied only the (Mold E2) horse, Edwin Hulbert said.

The Hartland Clydesdales were mounted on bookends or a base covered in red felt, and are found on a base with a clock or a zodiac ash tray. They are found in square signs, oval signs, and tall rectangular signs with a Plexiglas half globe. The signs were sold in pairs with the horse facing right in one and left in the other.

Posed in an accurate trot, the Hartland Clydesdales balance on two, broad feet, but they tip easily. That didn't matter since they were always fastened in place. I found one with two feet still screwed to a glass half-circle; it must have come from the half-globe sign. The screw holes are in the bottom of two feet, on one side of the body, or in both locations, depending on where the horse was fastened to the various styles of signs and souvenirs.

The 13.5" square signs were made by Lakeside Plastics of Minneapolis and Chicago. Edwin Hulbert said that Lakeside Plastics got most of the Budweiser sign-making (point-of-purchase display) business. Lakeside Plastics, Thomas A. Schutz Co., and Display Corporation International of Milwaukee were "display houses" that designed displays and won contracts to produce them. Hartland Plastics did "almost no work directly" for the breweries and other big companies, but was often subcontracted to produce displays or parts of them. Budweiser might have four to seven displays made in a year, and Hartland Plastics made the 7.5" high horses for "all the companies" that had those contracts, Edwin Hulbert said.

Big Eight Team. Signs using eight of the 7.5" high Clydesdales were made, too. They measure nearly six feet long. Ed Hulbert said that Lakeside Plastics made the sign (for Anheuser-Busch), but Hartland made the horses for the sign—and made them one horse at a time! The wagon and other parts of the sign were not made by Hartland, he said.

The story goes that, after 50,000 Clydesdales had already been made, Hartland got an order for 50,000 more. Ed Hulbert said, "A Minneapolis display house [Lakeside Plastics] ordered a beryllium copper mold from us to produce the big order they [got] from Busch. I spent several weeks trying to cast usable cavities, but finally failed, and we molded their order on our original electroplated mold." The failed effort was to make a mold of four 7.5" high Clydesdales as a unit. The larger mold could have been put on a bigger molding press and would have been a more efficient way to make the eight horses needed for each "Big Eight" team, Edwin Hulbert said in 1995. However, "the two halves wouldn't fit," and Hartland had to go back to using "an old copper mold" of the single horse. The failed four-horse mold was "made by casting die metal which was beryllium copper," he said.

For a while they [Lakeside Plastics] bought the horse from Hartland Plastics," but Lakeside may have made its own "steel molds of the [7.5"] horse later; I'm not sure," Hulbert said. Brewery sign horses of designs different than the Hartland Clydesdales were made in China or Hong Kong, he noted. He added that Hartland Plastics had no connection with the circular, eight-horse Budweiser hitch signs designed to hang from the ceiling. (Those signs used horses painted bay, but only 3.5" high.)

The Sculptor is Unconfirmed. Hartland's Alvar Bäckstrand was not the sculptor of the 7.5" Clydesdale, but he worked on it. He said in 1990, "We received the wood model from the beer company. From that wood model, I made a casting in Monel metal (a mixture of Cerro matrix and Cerro base with low melting point), and I made the left and right shells, and the "bead" parting line as on all other models, sharpened and improved the details, and removed all undercuts." (This was the only Hartland-produced horse that began as a wooden model.)

Anheuser-Busch spokesman Dan Reynolds could not say who the sculptor of the 7.5" high Clydesdale was without researching old records. Hartland's Roger Williams recalled sculpting some horses for breweries, but that might refer to the 2" Clydesdale that his wife said he made. (Also, Dan Reynolds said that the Jos. Schlitz brewery in Milwaukee used to have a model of a team and wagon; perhaps, it was sculpted by Mr. Williams.) Williams generally used clay for original models to be molded, and Mrs. Williams did not think the 7.5" high Clydesdale looked like his work.

In any event, Hartland did produce the mold, and molded and painted about 100,000 of the horses. They were molded in white styrene, and the body was painted orange-brown with black shadings, black mane and tail, and black harness with gold details; white (painted) stockings, white on the face, and red and white decorations in the mane and tail. (In 1967-1968, Hartland Plastics made 9" Thoroughbreds in this color, and called them, "Budweiser Brown.")

The Clydesdale Hartland made on the black base has flat rein loops; the raised loops were added sometime later, either by Hartland Plastics or Lakeside Plastics. The Big Eight hitch needed the rein loops to connect the reins from one horse to another, but many of the single horses in the beer signs have the raised loops, too. They may have been added when Ed Hulbert was trying to make a mold for four horses at a time.

The horse and dog models in this book are mostly between five and 40 years old. Few of them are still mint, but good care can prevent additional damage. Whether your purpose is to sell the model, maintain it as it is, or repaint (customize) it, the first step is cleaning.

Cleaning

Identify the Finish. Before cleaning the model, observe whether the model is painted all over, partly painted (such as only on the face, mane, tail, and lower legs, but not on the balance of the body), or entirely unpainted. Check rubbed (or broken) areas for the color of the underlying plastic. If no rubs are obvious on the hips, shoulders, etc., check the hoof edges, ear tips, etc. Painted models may be painted on the hoof bottoms, so that's not a reliable place to check. Also, some models have two layers of paint, and a light rub or scratch might only penetrate to the lower layer, rather than to the plastic. Note that unpainted models can be molded in white or colored plastic.

Unpainted models, and unpainted areas on partly-painted models can be cleaned fairly vigorously with a cotton swab, ordinary soap, and warm water. Check to see that the cleaning isn't scuffing the plastic. Be sure to run the swab over the fine lines in the mane and tail and between the ears.

Clean Gently. You need to be careful about cleaning painted areas since even gentle rubbing can remove the paint. I just glide a soggy, soapy Q-tip over the painted areas. Also, do not soak painted models or some of the top layer of paint "molecules" will come off.

A styrene horse with no paint on it can be cleaned fairly briskly, but check to see that the cleaning isn't scuffing the model's surface. Also, you should always grasp the model by its midsection; never hold it by the tail or leg because, with the tight grip needed for intense scrubbing, the model could break.

If you are not confident about cleaning the model, leave that job to the next owner.

No Harsh Cleansers. Do not use harsh cleansers, even on unpainted models, because they will remove the attractive, natural shine of the plastic. They will also make the plastic more brittle.

Removing Sticker Marks. To remove the marks left over by price stickers, use shampoo on a cotton swab.

Shake Them Dry. Most Hartlands were manufactured with a small hole on their underside, near the hind legs. If you submerse a model, it may take on water. So, after you're done washing the model, shake the water out, then stand it on a paper towel or in the dish drainer to dry.

General Maintenance

9" Five-Gaiters Need A Boost. Many older (1960s), 9" Five-gaiters will stand better if you place a film container cap, coin, jar cover, or plastic milk bottle cover under the left hind toe.

Avoid Heat. When storing, displaying, or transporting Hartlands, avoid extremes of temperature. Hartlands are mainly hollow, and their plastic bodies transmit heat to the air inside them. When the air inside the model warms up, it expands (takes up more space than before), and pushes on the inside walls of the model until a seam pops open to let some of the air out. Once popped open, seams can't be closed. (Heating the plastic to push the seam closed would also heat the air inside.) Collectors call this condition, "split seams," or "seam separation."

Besides the seams, there's another reason to not leave plastic models outside on a warm day. A half-hour in the sun can soften acetate plastic enough that a model's leg or ear tip could bend under slight pressure. I learned that the hard way with both a Breyer horse and a Hartland horse in the 1960s.

Being left in a parked car on a warm day, even with the windows cracked open, can be fatal to pets and small children, and is not recommended for Hartland statues, either. If you can't leave the window down, the trunk will be cooler.

Also, don't store Hartlands on a open shelf near a heater. Shelve them at least a foot away, in horizontal distance, from floor heaters. A horse on a high shelf that stands within 6" of a floor radiator will get warm because the heat goes mainly straight up. Attics, sheds, garages, and other places that get hot in summer are poor places to store Hartlands.

Some of the paint used for some batches of 1993-1994 Hartland horses was defective, and could turn sticky at about 82 degrees. Chances are that whatever damage was possible has already happened. However, each time the models are exposed to heat, they will get sticky again. If you don't have air conditioning, be sure to stand those models with nothing touching them, not even the string from a hang tag. The models are not sticky while the temperature is below about 82 degrees. Paola Groeber's test color models in styrene plastic and styrene models by Steven with a gloss coat added will be slightly sticky in warm weather, and dust will stick to them. (The problem of "tackiness" applies to far less than 1% of all Hartland horses; none of the 1960s-1970s models and most of the 1980s-1990s models are never sticky.)

Avoid Cold. In the cold, plastic is less flexible and can break more easily. Following an hour's drive in below freezing temperatures, a Breyer model's tail broke off this winter when the owner was unpacking it at a model horse show and bumped it only slightly.

Don't keep painted Hartlands near a humidifier, either. I had my humidifier on the floor a few winters ago, and a month or more later, a flake of paint started to lift on a Breyer horse on a bottom shelf about 15 feet from the humidifier. The paint was also starting to peel on the wall behind the humidifier. (I pinned a white plastic garbage bag to the white wall after I discovered the problem.) Breyer paint can peel, so why take chances with Hartlands?

Colors Can Fade. Sunlight is wonderful, but not for painted statuary. Don't put Hartland models on the window sill unless you want "a horse of a different color" one each side. In 2000, I saw a 7" Arab Foal in "yellow and gray." The gray paint on its left side had faded, and the plastic beneath it had turned yellow. The other side was normal.

Improvements

Gluing a broken leg or tail back on is the only improvement, besides careful cleaning, I recommend.

Repairs.To glue a leg or tail back on an acetate ("Tenite") horse, I use crystal clear household cement, which comes in small tubes of less than one fluid ounce. Brand names include Testors and Elmer's. I think "Model Glue" is specifically designed for styrene models. There is no remedy for "split seams," only prevention.

Don't "Touch Up." Do not attempt to touch up rubbed areas. The color won't match, and it will only make the model look worse (and decrease its value).

Shipping

Because of Internet auctions, probably more second-hand plastic horses are now shipped in a week than used to travel in an entire year, and many shippers are new to Hartland horses. Here are a few rules:

(1) For shipping, each horse should go into a plastic bag such as a soft, flexible bag from the grocery store. Don't wrap models directly in newspaper because it can come off on the model. The ink can even "merge" with the paint. Do not put two horses in the same bag because they will scratch or scuff each other.

(2) Some horses need wraps before going into the plastic bag. For adult horses in styrene plastic in the 7" series and smaller, it is a good idea to wrap some toilet paper or other soft material around the tail and hind legs to fill in the area between the tail and the body. This will help protect the tail from cracking or breaking off in shipping. For 9" and larger horses, it is a good idea to use extra, soft warp around the head to protect the ears and muzzle from getting rubs in transit. Travel is not a gentle process, and there will be a lot of motion in even the most well-packed box. Some collectors wrap the entire horse in toilet paper; that is a good idea for china horses, but not necessary for plastic horses.

(3) Next, wrap the bagged model in bubble wrap, tissue paper, more plastic bags, or newspaper, etc. Tape the material closed so it won't unwrap.

(4) Select a box large enough to leave extra room all around the model (at least an inch or two everywhere; more if your box isn't sturdy) for lots of cushioning material.

(5) Cushion the mummified (bagged and wrapped) model with packing peanuts, crumpled newspaper, crumpled paper bags, etc. The model should be surrounded on all sides by the cushion. If shipping more than one horse in the same box, always put heavier horses closer to the bottom, and lighter horses on top.

You might want to avoid shipping Hartlands in 90 degree weather (to avoid splitting the seams) or below freezing weather (in which a slight bump to the brittle model might result in breakage).

Be extra careful when packing the rare, sticky-finish model for shipping or storage. If there's a chance they'll get warm, don't wrap them in paper or cloth. (Avoid gift wrap tissue, facial tissue, paper toweling, toilet paper, cloth handkerchiefs, etc.) Don't try to dust the models while they are sticky, and it would be better to ship them in cool weather.

Customizing

Repainting. Repainting model horses for one's own pleasure is a branch of the model horse hobby. The object is not to retouch the model's color, but to change it to something new. Repainting usually reduces the (eventual) resale value of a model unless the work is outstanding. I recommend using paint you can soak off (acrylic or oil paint) so that if you change your mind, you can return the model to its original color. Enamel paint can't be removed without stripping all layers of paint off the model. Models for sale that have been repainted should always be described as such.

Customizing. Alterations to factory models can include adding paint, removing paint, or changing the shape of the statue by heating and bending part of it (such as the horse's leg), taking a part from one statue and gluing it to another, or adding material such as clay or putty to change the contour of the model. Adding a blaze or socks or a gloss finish to a model makes it a customized model.

Stripping. Some customizers remove all old paint before repainting a model. I've heard two stories lately of collectors trying in vain to strip the color from Hartland horses only to realize later that the color was molded in!

Restoring Copper Sorrels. I've shuddered at the results of people dabbing at the rubs on a model with a felt-tipped pen or paint brush, and that's what inspired my "do not retouch" rule. An overall restoration, on the other hand, could improve the model. Collector and artist Susan Bensema Young has a technique for restoring the color on copper sorrel 11" Saddlebreds. She uses Rub-n-Buff's copper color paint, available at art stores, to restore the top layer of paint for these models, which were originally painted in layers with gold over a red copper, almost magenta color. She advises thinning the paint with turpentine or other thinner that will not "disturb the Hartland paint," and brushing it on in thin layers. She said it's not an exact match, but it comes close, and it also works for the 9" copper sorrel Saddlebreds. It should work for the Tenite 7" Morgans, too. Most of the 7" Morgans are styrene, but Susan thinks that Rub-N-Buff won't harm styrene plastic.

Detecting Altered Models. Restored models should always be described as such. (Keep a notebook of your efforts so that if you eventually sell the model, its "pedigree" can be passed along.) For detecting an undocumented touch-up, Susan recommends comparing the paint surface on the most visible areas to the model's underside. If there is a difference, the model may have been retouched. Tip-offs of a retouch or repaint job are "anyplace where the surface finish is grainy, dull, or doesn't quite match the rest," and of course, "any part that isn't smooth like airbrush [painting], but shows brush marks."

Another sign of a non-factory finish can be, except for models noted under "Avoid Heat," a sticky finish. If the model's color is not a factory color, and the finish is sticky, it may be a repaint. According to the October, 1960, article in *Modern Plastics*, Hartland Plastics used paint from *Bee Chemical Co.*, Chicago, and *Wolverine Finishes Corp.*, Grand Rapids, Mich. The 1960s factory horses never had a sticky finish.

Styrene is Inflexible. While both acetate ("Tenite") and styrene are thermoplastics, meaning that they can be (commercially) reheated and re-molded, styrene is much less flexible, and customizers avoid it in favor of Tenite.

Heating Acetate ("Tenite") Plastic. Whether you're trying to customize a model, help a 9" Five-Gaiter stand better, or straighten a crooked, heat-warped leg, here are some things to keep in mind. "Coffee hot" water can be used on white plastic, but will turn colored plastic lighter, and it will never be the same. The 1960s and 1970s models are so old that their plastic is inflexible. If you move a leg, it may go back to its original position within minutes. (Plastic has "memory"; old plastic will revert to its long-held shape.) Moving a leg may still work for the 1980s-1990s models, but there is a risk that the paint could bleed. Heating works well with classic-sized Breyers that are a year or two old; their plastic is still very flexible and their legs are thin.

Moving A Leg. Paola Groeber said that she sealed her Tenite models with a semi-gloss finish that prevents the paint from "bleeding out" if the model is placed in "coffee hot" water in order to reposition a leg. She cautioned that Tenite models painted by Steven Mfg. may not be sealed, so don't try that on them. (Some models by her and Steven are the same color, so unless you are sure of the manufacturer, don't heat them.)

Paola's technique for repositioning the 9" Five-Gaiter's left hind hoof is to, while wearing a mitt, hold the model in a pot of steaming, but not boiling, water, with the water level up to the horse's hips. Do not touch the model to the bottom or sides of the pot. Count to 30, then remove the model from the pot and pull the leg back, then forward and down. Rinse it immediately in cold water. Stand it on the counter to see whether the model now balances or needs more work. Reheat and repeat if necessary. The model will get hot, so don't hold it barehanded. Instead of holding the model in a pot, I've poured the hot water into a tall, straight-sided glass or mug and held the model there.

To straighten a leg on a Hartland molded in colored, acetate plastic (such as a woodcut), Paola recommends cooler water (with no bubbles) and doing it in stages. She said she soaked a 9" Tennessee Walker's leg in the tea water at a restaurant during most of the meal, moved the leg, and then stuck it in the ice water to "set" the new position. (Neither she nor I are responsible for strange looks or other results of readers following the care and repair instructions in this chapter.)

Resin Models. Paola also said that the hot water method works on the cold-cast resin models she painted and sold since they have a plastic resin base. Should they get a bent leg or hoof, it can be straightened without breaking the model or damaging the paint job although "it takes a little longer and you have to be very careful." She went on to say that the resin cast models are "very popular because they are artistic, but also very durable. They don't break as easily as a porcelain model, but they will break if dropped from a high place onto a hard surface."

The story of Hartland horses is complicated and compelling. They have had a succession of manufacturers, and for a ten-year stretch, they weren't made at all. If they reflect an artistic, imaginative, and fun-loving touch, it should come as no surprise: The two people most responsible for their manufacture were a musician and a magician.

A Summary of Hartland Model Horse History

The First Three Hartland Horse Companies. Hartland models began in Hartland, Wisconsin, just west of Milwaukee. The first horse by Hartland Plastics, Inc., was supplied to Mastercrafters Clock and Radio Company, Chicago, for a mantle clock in the late 1940s. (Incidentally, Breyer's first horse, the Western Horse, was a copy of it.) Called the Large Western Champ, the horse was named after Paul Champion, Hartland's national sales manager. He and especially Robert McGuire—a lawyer, marketing executive, and horse lover—selected the breeds, colors, and sizes of animal figurines, and named them. In the 1950s and early 1960s, Hartland set the pace with its high-quality, plastic, horse-and-rider models. In the balance of the 1960s, its individual horses and dogs and horse families of various breeds gave animal lovers a nice selection to choose from.

Then, Hartlands left their first home. Things go in cycles. In the late 1960s, the Baby Boomers were outgrowing their need for plastic horses, and people were buying fewer toys. In 1969 Hartland Plastics discontinued the toy part of its business and sold those molds to Strombecker Corp. of Chicago. Strombecker's Durant Plastics division in Durant, Oklahoma, made only a portion of the horse and dog line between 1970 until fall of 1973, when the Arab Oil Embargo caused the price of plastic to go up. Styrene and other plastics are made of oil.

In the 1970s, model horse collecting emerged as an organized hobby with newsletters, clubs and photo shows by mail, and gatherings in person, called live shows. But whether organized or not, those who were aware of Hartland models and frustrated that stores and mail order companies no longer sold them, endured a long wait. Early in 1979, the news came from collector Bettye Brown that Steven Manufacturing Company of Hermann, Missouri, had purchased the Hartland molds. She and Linda Walter's *The Model Horse Shower's Journal,* urged collectors to write to the company. The letter-writing campaign brought no immediate results, though. In 1982, when I was finishing the text for the first edition of this book, I checked with Steven Mfg. to see what their plans were. Steven's reply was that whether the molds would be returned to use would depend on their condition and the cost to refurbish them. According to Paola Groeber (in the July/August 1985 *Model Horse Gazette),* my inquiry "stimulated enough interest to check the molds out and run a limited number for a test." At that time, Paola Groeber worked for Steven in sales.

Steven made Hartland horses from 1983 through 1994, and had two different owners during that time. From March 1992-March 1995, Steven Mfg. and its Hartland division were owned by a group of investors, and some of the models did not match the previously high standards. In March of 1995, Steven and Hartland were bought back by Mr. Bev W. Taylor, who had headed the successful Hartland production from 1983 through 1991. Although the company planned to eventually resume Hartland production and add seven new equine shapes that had been commissioned prior to 1992, those things did not happen. In late February of 1997, Mr. Taylor sold the company—to David Segal—and retired for the second time. The Hartland molds are still dormant.

The Fourth Manufacturer. Many fine models were made by Steven from 1983-1994, but there was also a fourth manufacturer of Hartland horses during part of that time, Paola Groeber. Paola started a small, spare-time business, Hartland Collectables, Inc. ("HCI"), in 1985. At first, she was just a mail-order seller of Steven's Hartland models, but in 1987, she became a manufacturer in her own right, obtaining molded bodies from Steven's molding division, painting them in original colors (during evenings and weekends, in the basement of her home), and marketing them independently of Steven Mfg. She advertised in model horse journals and sold by mail and at model horse shows. Paola said that at the first BreyerFest (in 1990), her carload of models sold out on the first day. The product was a success, and she had trouble keeping up with orders. Then, despite Paola's protest, Steven stopped molding horses in Missouri in favor of having the work done in China for a large order from JCPenney. With Paola's supply of 9" series model bodies brought to an end, she closed HCI at the end of 1990. She filled remaining orders in 1991, and painted and sold a handful

of resin horses (cold cast models) in 1992. In December 1994, fall of 1999, and spring of 2000, she sold additional models left over from her manufacturing days.

The Future? Hartlands have been through a lot. Two severe floods in the midwest, in fall of 1986 and summer of 1993, caused setbacks to Hartland at the Steven plant in Missouri, but Hartland production resumed both times before ending in 1994. The molds are old, but most are in usable condition, and molds can always be repaired. Whether Hartland horses will be made again, however, remains to be seen.

One thing, for sure, is that model horses are no longer viewed as just children's toys, something that a young person should outgrow. Since sometime in the 1980s, collecting just about anything, including old toys and models, has been viewed as a respectable, adult pastime, rather than a hallmark of eccentricity. Model horses became engulfed in the collectibles craze in the early 1990s.

Two Other Hartland Manufacturers. Between the mid-1980s and late 1992, two additional companies (they could be called the fifth and sixth Hartland manufacturers) re-released Hartland baseball player statuettes only. Both companies, Bill Alley's Hartland Plastics, and USA Hartland, which was part of Case-Dunlap Enterprises, were located in Dallas, Texas. A summary of them is found in *Hartland Horsemen* (1999) and *Hartland Horsemen & Gunfighters* (1998). My article in *Toy Collector and Price Guide,* September 1995, includes a longer account. Information on the original sports statues was published in the 1991-1995 editions of *Hartland Horses and Riders,* and in *Hartland Market,* August 1995; both include my interviews with sculptor Alvar Bäckstrand, including his account of showing a clay, model-in-progress to a baseball player. Additional information will appear in a future book.

Hartland Plastics: The Original Hartland Company

The story of Hartland model horses goes back to 1939 and the village of Hartland, Wisconsin. Hartland is about 30 miles west of Milwaukee, in the lakes region of Waukesha county, in southeastern Wisconsin. A hundred years ago, Waukesha county was known for its health spas, resorts, and lake estates: it was a summer playground for people from Chicago and around the world. It was primarily an agricultural area, but with more than its share of horse farms. The Pabst brewing family bred draft horses there, and Montgomery Ward, the catalog king, raised Hackneys that won prizes in four-in-hand coach classes. While some of the farmland has, of course, given way to development, there are still stables in the Hartland and nearby Oconomowoc (oh-CON-oh-mow-walk) areas, though now mainly for Saddlebreds and hunters. In 1940, the population of Hartland was 998. In 1990, it was 7,735, but Hartland's downtown is still only a few square blocks, with the area's affluent heritage visible in upscale shops and restaurants. Here, in 1939, the forerunner of Hartland Plastics began in a building that once housed a harness shop.

Edward and Iola Walter. Edward Walter was born in Chicago, February 14, 1895, schooled in Detroit, served in the Navy in World War I, and then worked for his father's manufacturing business in Kitchener, Ontario, Canada. He and Iola White married in Toledo, Ohio, in 1930. By the time they started business in Hartland, Ed Walter had acquired experience in various phases of the new industry of plastics manufacturing. He was known as an excellent mechanic.

The Walters' first products were candle molds and related items. At first, Iola operated the shop alone, at 132 (later renumbered 140) Cottonwood Ave. The building, which no longer stands, was three doors south of Max Meier's restaurant, the Hartland Inn, at the hub of downtown Hartland. After her husband joined her in the business, they moved it to 112 W. Capitol Drive, at the corner of Capitol Drive and Cottonwood Ave., kitty-corner from the restaurant. There, Ed Walter converted the Dodge garage and service station building into a factory. Their major product in the early 1940s was plastic heels for military shoes since World War II was in progress.

The Walters Gain A Partner. In 1943, Edward and Iola Walter incorporated their business as Electroforming Co., for the purpose of manufacturing, purchasing, and selling plastic, metal, and other materials. The name was changed to Hartland Plastics, Inc., on March 19, 1946. The same year, they added a 55' x 80' extension to the west side of the plant, and in 1948, a two-story addition measuring 40' x 111', followed by a third addition in 1951. (The building, now subdivided into storefronts, still stands.) A contributing factor in the success and growth of the company was the Walters' acquisition of a financial mentor and partner, Charles Caestecker, of Chicago and Kenilworth, Illinois. Charles Caestecker owned American Molded Products Company, Chicago, which was later known by the name of one of its divisions, Amerline Corp. Caestecker, a professional musician as a young man, had studied violin in Europe. However, his additional interest in business led him to start American Molded Products. He became part owner of Hartland Plastics sometime in the 1940s.

Roger Williams. Hartland hired the first of its two sculptors, Roger T. Williams, in 1944 or 1945. After the war, Hartland Plastics produced ornate, wall decor: brackets and candelabra, and ornamental mirror frames and picture frames for Silas Levy's Cameo Miniatures of New York. Roger Williams did the carvings for the frames. Hartland also made birdhouses, started its line of religious statuary, and did work for the Mastercrafters Clock and Radio Company, 216 N. Clinton St., Chicago. Mastercrafters made grandfather clocks and mantle clocks, and Roger Williams sculpted the plastic figurine for the "girl in a swing" clock, and a clock horse that launched two brands of plastic horses.

The Clock Horse. Hartland produced the horse to stand over the Mastercrafters Clock in about 1947 or 1948. In 1948, Hartland Plastics offered "a complete line of religious articles," which were plastic statuettes. By the time the clock horses were being produced, the Hartland religious line was already occupying so much of the company's attention that Ed Walter decided to discontinue supplying horses to Mastercrafters. Some time after that, Mastercrafters turned to Breyer Molding Company, Chicago, and Breyer filled the order with a clock horse that was a close copy of the Hartland clock horse. Later, Hartland and Breyer each started their own lines of model horses with the respective design they had first used for Mastercrafters. (More details on the clock horse are found in *Hartland Horsemen*.)

Paul Champion, Robert McGuire, and Edwin Hulbert. In 1948, when the plant employed about 18 or 20 people, Hartland Plastics hired Paul Champion as manager of sales and advertising. Robert McGuire, an attorney, was hired in 1950. He was in charge of personnel, purchasing, and public relations, and was also a marketing executive.

In May of 1950, Edward Walter was treated for an illness that later turned out to be Hodgkin's disease. Edwin Hulbert said that by the summer of 1951, Walter was still up and around, but not entirely well, and persuaded him to come to work at Hartland Plastics. Edwin Hulbert, who with his father, Ed Hulbert, Sr., owned Hulbert Engineering Corp., had known Ed Walter since the early 1940s. (Hulbert Engineering designed and built machinery and tooling for converting rigid, sheet plastic into packaging and displays, but it was closed after Ed Hulbert, Sr., died in October 1951.) Ed Hulbert began as general plant manager of Hartland Plastics August 1, 1951. By fall, Walter was bedridden at his home on Pewaukee Lake. With frequent visits from Paul Champion and Robert McGuire, he continued to run the business from home. When Edward Walter died December 28, 1951, he was only 56.

Hulbert said that in 1952, Charles Caestecker and the attorney for the Walters decided that "Robert McGuire and I could handle the plant and office, and Paul Champion could handle sales. We could make a go of it." A friend of Ed Walter's from Cutler-Hammer, Inc., a Milwaukee manufacturer, helped install more efficient equipment, and additional help in upgrading the plating department (a key step in creating molds) came from Plating Engineering Company, Milwaukee, Ed Hulbert recalled. With Charles Caestecker as president, Iola Walter continued as a corporate officer (secretary-treasurer) of Hartland Plastics for two more years before selling her half of the company to Caestecker.

Brand Names. A manufacturers' directory for 1952-1953 and 1954 listed the Hartland Plastics' brand names as, "Pearl Glow," "Iolite," and "Diamond 'I'." Edward Walter had evidently named the brands after his wife. Since the directories gathered their data up to a year in advance, these brand names were probably being used in 1951 through 1953. They were probably used earlier, also, but the 1950-1951 directory did not list brand names, and Hartland Plastics did not appear in earlier editions. In the 1955-1973 directories, the only brand name given is "Hartland"; which means the Hartland brand was in use from 1954 through 1972. No brand names were listed for Hartland Plastics in the 1974-1978 directories. By then, the company was engaged entirely in molding industrial components and custom molding (for clients).

Charles Caestecker. From 1954 until 1965, Charles Caestecker and his wife, Marie, and son, Thomas, were the sole owners of Hartland Plastics, and Charles Caestecker was president through 1970. Most of Hartland Plastics' models were made while Charles Caestecker owned the company. It couldn't have had a better owner; everyone I talked to spoke highly of

him. Hartland Plastics never had a union, and didn't need one. As one former employee said, "Charley Caestecker was a nice guy."

Alvar Bäckstrand. Roger Williams did all of the sculpture before Alvar Bäckstrand arrived in 1956. Then, they worked concurrently until 1968, when Roger Williams retired at age 65. Alvar Bäckstrand continued with Hartland Plastics until he retired in 1977.

The Three "Idea Men." An article on Hartland religious statues in *Life* magazine, April 20, 1959, called Ed Hulbert, Robert McGuire, and Paul Champion "idea men" of Hartland Plastics. In the 1990s, Paul Champion explained that Hartland Plastics' product line decisions were made during restaurant lunches attended by himself, Robert McGuire, Edwin Hulbert, and Edward Walter or Charles Caestecker. Edwin Hulbert added that they held conferences with the sculptors, Roger Williams and Alvar Bäckstrand. The Hartland baseball statues (1958-1963) were the idea of Thomas E. Caestecker, Charles' son. Hartland Plastics was a relatively small company guided by the ideas of a handful of people.

Supporting Staff. Hartland Plastics' two sculptors and supporting staff were exceptionally good at making figurines, at making the ideas tangible. In the 1950s, the company typically employed 30 men and 70 women, including many young women who worked in the painting department. Paul Champion recalled between 10 and 30 painters working on religious figures, horses, or riders at any given time.

In 1955, Hartland Plastics moved to its final location in Hartland, 340 Maple Ave. An employee who also owned a fur farm, Jerome Delsman, said that they "rented a hay dryer so the guys could lay block" for it. By 1963, employment was 140 and a 23,000 square foot addition (making 80,000 square feet in all) provided more production and warehouse space. The building on Capitol Drive was kept as a warehouse, along with a warehouse in Oconomowoc.

During 1964-1967, 40 men and 110 women worked for the company. From 1968-1970, it was 30 men and 150 women. In 1971-1972, the totals were 50 men and 150 women. Many of the women operated molding presses, glued models together, or packaged finished models, but the majority were spray painters. From 1973-1977, the work force statistic was 200 since a gender breakdown was no longer politically correct. (Numbers listed here were adjusted one year backward from the manufacturers' directory cover date since data was evidently gathered up to a year in advance.) Hartland Plastics was supposed to have employed up to 260 at peak times.

The Horsemen Series. While religious statues continued to be a large part of Hartland's figurine line throughout the 1950s, the second major line of models began in 1952: horse-and-rider sets depicting historic characters and/or the stars of television westerns that were popular from the mid-1950s to the early 1960s. Hartland made 48 molded shapes of riders in four sizes (scales). With some of them painted in more than one color scheme, there were 63 different riders, not counting a few that were sold unpainted. Almost all of the riders had a differently-shaped hat, and many had a unique saddle, weapon, or other accessory. Hartland created 20 different shapes (11 basic shapes plus variations) of horses to accommodate the riders, and most came in many colors.

The Horsemen series was successful. As Paul Champion said in 1997, "TV sold [Hartland's western] stuff. We tagged along. When TV dropped cowboys, we died." However, the western models paved the way for the breed series horses and other animal figurines that followed.

Breed Series Horses. Paul Champion said that in the toy business, there was always a need to come up with something new and different, so he thought of making horse families to replace the rider series. The 5" and 7" Arabian families appeared in 1961, followed by more horse families, 9" horses, Tinymite horses and dogs, and 11" series horses. Paul Champion wrote most of the text for the Hartland catalogs and packages, and would consult with Robert McGuire on horse breeds. They both thought of the names—for horses that were named. The Tinymite horses and dogs were also Paul Champion's idea. In 1996, he lamented that the dogs did not sell very well, but that's partly midwestern modesty. (Living through extremes of temperature diminishes the tendency to boast or exaggerate.)

In 1995, Thomas Caestecker said of Hartland Plastics, "We were willing to try anything. Our mold costs were not excessive." He added that an important part of his father's [and Hartland Plastics'] success was the ability to recover from mistakes. As he said in 1999, the objective at Hartland Plastics was to "think of something new, [figure out] how to make it better, and make a living, but something beyond that. [They] had fun doing what they were doing."

Robert McGuire loved horses, and both he and his father, a horse trainer, had been in the cavalry (his father, in World War I, and he, in the cavalry equivalent of ROTC for four summers during high school, Rita McGuire, his wife, said). There was a horse collar on the office wall at Hartland Plastics, and Robert McGuire once joked that they should call the company "Horse Collar Plastics," Paul Champion recalled. (In fact, the employee newsletter, published by Edwin Hulbert, was called "Looking Through the Horse Collar.") For a time, Robert McGuire wore boots and a cowboy hat to work. He wanted to give up his office occupation for a ranch in Idaho—he'd even made down payments on the land—but it didn't work out due to a change in fortune.

Owned by Revlon. In 1964, Hartland Plastics became a division of Amerline Corp., which Charles Caestecker also owned, and in November 1964, he sold Amerline (including Hartland Plastics) to Revlon, Inc., taking payment in Revlon stock valued at about $18 million. The sale was final in spring of 1965. At that time, Hartland Plastics started making containers for cosmetics (compacts and lipstick cases) and Revlon shampoo; packaging for Avon; bottle caps for Johnson Wax and Esquire Shoe Polish, and many other consumer product manufacturers, Edwin Hulbert said. This work included doing precision, hot stampings of brand names or decorations. The production of point-of-purchase displays, such as display cases for Timex watches and beer and soft drink signs, also began in earnest. The manufacturers' directory for 1965 lists the company's activities as, "custom molding, decorating [meaning, painting], mold-making, point-of-purchase advertising displays, toys, statuary, and figurines, [and] industrial moldings." In the 1965 and 1966 directories, Hartland Plastics is listed as a division of Amerline Corp., not of Revlon. (The directory for 1963 was not available.)

Amerline Corp. was one of the prominent plastics molders in the United States. It made tape recorder reels, plastic parts for AT & T, containers and reels for IBM computers, and coil bobbins for wire for the electrical industry. For years, Hartland Plastics had 2-3 machines working full time just producing coil bobbins for Amerline Corp. Amerline, itself, had 15-to-20 or more machines making coil bobbins. Amerline also had a patent on a plastic, water closet float. The steady business in non-glamorous products allowed Hartland Plastics the luxury—and fun—of making toys and other creative, decorative products.

Except for adding some new types of products, not much changed while Hartland was part of Revlon. Charles Caestecker was still president; "idea men" Champion, McGuire, and Hulbert, and sculptors Alvar Bäckstrand and Roger Williams were still at work. Hans Seuthe was first-shift supervisor of the molding department as he had been since 1949. Jerome Delsman was second and third shift molding supervisor. John Nicholas was still in charge of painting and assembly. Horse and other animal production continued, but new horse and dog molds that were created during 1965-1968 are not marked with the Hartland name or any brand mark. Some of them have the typical "assembly letters," though.

Contrary to rumor, no molds were ever "taken to the dump" on account of being part of Revlon or for any other reason. Edwin Hulbert said so in 1995, and in 1992, Tim Ford, a then-owner of Steven Manufacturing Company (the third Hartland company) told me that Steven Mfg. possessed nearly all of the old Hartland molds.

Model Horse News. In 1965 or 1966, the first model horse newsletter began, and it was connected with Hartland Plastics. The *Model Horse News* was published by Mission House, a mail-order seller of model horses. The connection with Hartland is that Mission House—later renamed Mission Supply House—has been owned continuously by relatives of Rita McGuire—the wife of Hartland Plastics' Robert McGuire. Mrs. McGuire, a teacher, also owned it, herself.

Her parents, Albert and Ruth Rosier, started the company in Floral City, Florida, in 1954. It sold Hartland models, but by the 1960s, included Breyer horses and other toys and dolls. Ownership then passed to her brother, John Rosier, and his wife, LaVerne, who published *Model Horse News*. Then, Mrs. McGuire owned it from 1982-1992. (Robert McGuire died in 1990.) Since June of 1992, it's been run by Rita's younger brother, Joseph Rosier, and his family in Lake Mary, Florida.

Model Horse News was published bi-monthly for about a year, Rita McGuire said. I've seen the January-February, May-June, July-August, and September-October 1966 issues. Each was four pages with articles on breed characteristics, letters from collectors, a "pen pals wanted" column, photos of Hartland and Breyer horses, and contests. Typical contest instructions were, "We want you to write in 50 words or less why you like the Belgian Draft Horse. There aren't too many of them around. Content of entry is important, but try to be neat." Three winners received the prize of a Breyer Belgian model in dapple, chestnut, or gray and white. Feature articles included, "The Gaited Saddlebred: Born Beautiful," "Western Horses: Wonderful Wranglers," "Belgians: Big and Beautiful," and "Thoroughbreds: King of the Track." It was a delightful newsletter well suited to teen and pre-teen horse lovers.

Marketing. Hartland models were advertised in newspapers and magazines, were found in dime and department stores, in mass-market catalogs (of Sears, Roebuck & Co., JCPenney, and others), and in specialty catalogs of toy, novelty, or model horse sellers. The products were featured modestly; Hartland never advertised on TV, and "TV controls a lot," said Paul Champion, national sales manager of Hartland Plastics from 1948-1966. TV inspires purchasing aspirations.

In the 1950s and 1960s, "Toys were a risky business." He remembered a trade show at which the originators of Barbie dolls had only one or two sample dolls, and remembered feeling sorry for the man at the Hoola-Hoop booth before that phenomenon caught on. He saw Etch-A-Sketch and Play Doh get started and become popular at the same time Hartland's toys were. Many plastics molding companies were busy in the 1950s and 1960s.

Hartland Plastics had 10 regional sales reps, listed in the 1968 horse catalog: Murray Gilbert Assoc., Inc., New York; Coursey & King, Inc., Atlanta, Georgia; Conway-Carey Co., Omaha, Nebraska; Cecil Swinney, Jacksonville, Texas; Witz-Knight, Inc., Chicago, Illinois; Will Schultz & Co., Cincinnati, Ohio; Louis Anderson, Minneapolis, Minnesota; Jim Wolf, Tacoma, Washington; Philip A. Johnson Co., Denver, Colorado; and William Marsh, Los Angeles, California. To that list, Champion added Warren Thomas in Ohio and Felix Kapp in Minneapolis.

Rita McGuire, Robert's wife, explained that in those days, there were also toy jobbers, companies that bought large orders of toys at (manufacturers') toy shows, warehoused the toys, and then called on stores and other retail distributors which then sold the toys to the public. Milwaukee, alone, had three toy jobbers, Champion said: M.W. Kasch was one of them. Wisconsin DeLuxe Company was another.

Taking Flight. Paul Champion frequently flew his private plane to sales appointments and conventions around the country. He laughed to recall that, one morning with the plane on auto pilot, he looked up from a book directly into the sunrise and realized he wasn't heading south to Louisville, Kentucky, as he thought!

On direction in life, Paul Champion said in 1990 that, when he addresses groups of young people, he tells them, "You are the only one who'll take care of you. Make decisions and take a chance." He came from the Westville/Georgetown area south of Danville, Illinois, and knew both good and bad economic times. He came to Wisconsin after his father was killed in a mine accident. He played such a vital role at Hartland Plastics that after he left, it was difficult to find a similar position in another company right away, but he bounced back. He commented that, "Affluence hasn't helped human standards."

A Change of Direction. In 1967, Charles Caestecker, Robert McGuire, and Edwin Hulbert bought Hartland Plastics (but not Amerline Corp.) back from Revlon. McGuire and Hulbert each borrowed heavily in order to do so. (Paul Champion had quit over a marketing disagreement in 1966.) The market for plastic horses was declining, and it was the last year that new horse shapes were created. After the primary animal sculptor, Roger Williams, retired in 1968, the company went in a new direction, Paul Champion said. It abandoned statuary in favor of functional products such as sculptured cabinets, which Alvar Bäckstrand was proficient in designing. Champion said that when he visited the plant in 1968, there was a display case of horses, but as a product line, they were already "dead."

Four, Unmade Horses. Robert McGuire said in 1982 that the Hartland Toy Line, including horses, was discontinued in 1968 and slowly phased out over 1968 and 1969. The Spring 1969 catalog was the last one to feature model horses. Unfortunately for collectors, four more 11" series horses that had been designed never had a chance to be put into production. They got so far as the metal model stage, but their molds were not finished. The metal models, pictured in this book, are owned by members of the Robert & Rita McGuire family.

Toy Line Sold. In July of 1969, the Hartland toy division was sold to Strombecker Corp., Chicago. Hartland Plastics' production manager, John Nicholas, (who had been in charge of assembly and painting for 20 years before that) was quoted in *The Waukesha Freeman*, December 23, 1969, as saying that no toys were manufactured in 1969, and that the complete stock of toys on hand was sold by November 1969. Over 40 tons of molds were moved from the Hartland plant to Durant Plastics, a Strombecker subsidiary in Durant, Oklahoma. In the four years prior to that, toys and cosmetics containers had been one-third of Hartland's production. Another third had been, "custom molding, painting, finishing, hot stamping, mold making equipment, industrial parts and components, and point-of-purchase displays." The final third was miscellaneous, industrial products or parts, the article said.

"We are in our glory doing this point-of-purchase work," John Nicholas was quoted as saying. Using the example of a beer display sign, he explained that the firm made the mold, decorated the sign, assembled it, and shipped it. With toys no longer being made, the company actually added more employees, for a total of 260. "Now we can concentrate on the things we should be doing," Nicholas said.

The cabinets did not sell so well as expected, however, and in 1970, Charles Caestecker, Robert McGuire, and Edwin Hulbert sold Hartland Plastics to Republic Pictures Co., taking payment in Republic Pictures stock. According to Edwin Hulbert, Republic Pictures "got into financial difficulties almost immediately." Its stock fell to one-quarter of its previous level, and the trio took a heavy loss. Charles Caestecker had reached retirement age, but McGuire and Hulbert were heavily in debt (because they still owed on their loans), and continued to work at Hartland Plastics.

Republic Pictures then sold Hartland Plastics to Familian Corp., a wholesale plumbing supply company in Van Nuys, California, in 1971, and Hartland received assignments to make drain, waste, and vent pipe fittings and valves and fittings for lawn sprinklers. In April 1972, most of the management personnel, including McGuire and Hulbert, were abruptly fired. (They were still listed in the 1973 manufacturers' directory, which is evidence that directory data was gathered at least nine months in advance of the cover date.) McGuire and Hulbert were notified by a phone call from California. As sculptor Alvar Bäckstrand said in 1990:

Things were going on under the surface that Ed Hulbert and Bob McGuire did not know, nor anybody knew, except [an engineer Familian Corp. had hired], and possibly one or two more.

One morning I arrived at the back parking lot at the same time as Bob McGuire. He looked more serious than usual, and as we walked into the factory, he said, 'Well, Al, this is the last day in the old castle.'

I laughed and said, 'I hope not.'

He said, 'Maybe you don't know. Ed [Hulbert] and I were fired yesterday. They gave us a couple of hours to get out of the factory.'

On the assignment of Familian Corp., Hartland Plastics had spent $100,000 to make the steel mold for a newly-patented valve design for lawn sprinklers. After the valve parts proved faulty, Edwin Hulbert refused to let Hartland spend more time on it. The engineer convinced the owners that the fault was with management. Ed Hulbert said that after he and the others were gone, the firm never did get the valve to work. The problem was in the design, not in the manufacturing.

The events of 1972 were, as Alvar Bäckstrand said, "the beginning of the end of Hartland Plastics." Edwin Hulbert returned to Watertown. Robert McGuire had to give up the dream of a ranch in Idaho. Instead, he bought the old Hartland plant on Capitol Drive, and remodeled it into shops and offices. For his efforts, Robert McGuire was named Hartland's "Man of the Year" in 1975. (The building has now been a shopping center for 25 years.)

Though its days of making figurines were years past, Hartland Plastics remained in Wisconsin until 1978. In October, 1976, Familian Corp. sold Hartland Plastics to Plastech Research of Rush City, Minnesota for $562,500. The sale included the plant and 12 acres at 340 Maple Ave, machinery and equipment, inventory, and the name. At that point, Hartland Plastics was almost bankrupt.

The change of ownership was supposed to be beneficial, but resulted in layoffs. Molding operations ceased in August, 1977. Secondary work such as painting continued until the plant in Hartland closed for good on June 30, 1978. Its work was supposed to be transferred to Minnesota, but by that time, there wasn't much work left to transfer. When I contacted Plastech Research in 1980, the Hartland division in Rush City was making such things as nose cones for snowmobiles, but it closed later in that decade. In 1999, Plastech owner Dennis Frandsen said he wished he would have kept the plant in Hartland going. (Since 1978, the building on Maple Ave. has been home to a series of manufacturers, some of whom were once clients of Hartland Plastics.)

Work (or Ownership) Dates for key people of Hartland Plastics are:
Iola (White) Walter, founder and owner from 1939-1953.
Edward Walter, founder and owner from 1939-1951.
Charles E. Caestecker, owner from the 1940s-1965 and 1967-1970; also, president from 1965-1967.
Roger T. Williams, sculptor, 1944 or 1945-1968, and free-lance work for Hartland until 1970.
Paul E. Champion, national sales manager, 1948-1966.
John Nicholas, in charge of finishing (assembly and painting), and

from about 1969, was production manager, 1949 (or earlier)-1972.
Hans Seuthe, first-shift molding department foreman, 1949-1977.
Robert B. McGuire, attorney and marketing executive, 1950-1972; also, owner, 1967-1970.
Edwin F. Hulbert, general plant manager, 1951-1972; also, owner, 1967-1970.
Alvar Bäckstrand, sculptor, 1956-1977.

Although its fortunes sometimes changed rapidly, the original Hartland company lasted four decades, and we're still enjoying its products.

Durant/Strombecker: The Second Hartland Company

Strombecker Corp. of Chicago bought the Hartland toy line tooling (molds) in July 1969. The purchase included some of the leftover supply of plastic model horses that were still on hand at the time. Strombecker then marketed Hartland horses under its own name from 1970-1973, when it stopped due to declining sales. The Arab Oil Embargo in fall of 1973 may have had an impact, also, since the price of oil affects the plastics industry. Thus, the fate of plastic horses rises and falls with the state of the world economy. Horse collectors have a special reason to read the business pages.

In 2000, Strombecker officials said that they used the Hartland Plastics "catalogs" for between six months and two years. During that time, some old models that had been produced in the 1960s by Hartland Plastics were sold by Strombecker in packages that read, "1970 Strombecker Circle H Durant, Oklahoma." As the old supply of various horses sold out, Strombecker replenished the stock with new models made at its Durant, Oklahoma, subsidiary, Durant Plastics. Circle H was a brand name used by Strombecker.

I regret to say that the actual models produced by Durant were not, in many cases, molded as neatly or painted as attractively as the 1960s horses. For example, the Durant catalogs show a Tinymite Arabian painted brown with flaxen points, like the 1960s model. However, The Tinymite Arabian I bought at Snyder's Drugs in 1975 in the 1970 Strombecker Circle H Durant package is painted orange with flaxen points. Durant produced the horses and dogs as toys, rather than collector models. Where the 1960s horses in the 5" and 7" sizes usually have shading on each side of the face, the Durant-produced models do not. The Durant-produced horses have the same mold-mark (or lack thereof) on the inside leg of the model as the 1960s horses.

The leftover 1960s horses sold in the 1970s by Strombecker may have included horses that Strombecker did not put in its regular 1970-1973 catalogs. I say this because in 1974, I wrote down the Strombecker address found on a friend's Hartland horse catalog that I'm sure included a picture of the 7" Arabian Stallion, with its distinctive, left hind foot position. The 7" Arab Stallion is not in the 1970-1973 catalogs from Durant/Strombecker, though.

Another possibility is that the catalog was from the 1960s, and Circle H was a distributor of 1960s models by Hartland Plastics. Hartland's Robert McGuire had said that, "Circle H was used on all Hartland products that did not carry the 'Hartland Plastics' copyright I.D." He may have been referring to the horse and dog molds that originated during 1965-1968, because they are not marked "Hartland." Unfortunately, we can't get further clarification on this from him.

In 2000, officials at Strombecker said that it no longer uses the Circle H brand name, but toys are still manufactured at Durant Plastics, which (in February 2000) employs 60 people. The city of Durant, Oklahoma (population: 12,929), is near the southern boundary of the state, and is home to Southeastern Oklahoma State University.

Steven and Paola: The Third and Fourth Hartland Horse Companies

In the late 1970s (perhaps, July of 1978), Steven Manufacturing Company, a toy manufacturer in Hermann, Missouri, bought the Hartland molds from Strombecker Corp. and in 1983 became the third Hartland manufacturer. Steven Mfg. (which, until 1992, was a division of Handi-Pac, Inc.), produced horses from 1983-1994. It also made possible the small business owned by Paola Groeber that is counted as the fourth manufacturer of Hartland horses. The stories of Steven and Paola are intertwined from 1985-1990. As of January 2000, Steven is still the current owner of the Hartland name and tooling. The horses produced in Missouri are fully half of the story of Hartland's breed series horses.

Steven Mfg. Co. About 1940, a boy named Steven Zemelman wanted a kaleidoscope, so his father, Roscoe, built one himself, and eventually started a company in St. Louis named after his son. Steven Manufacturing Company began making kaleidoscopes in 1942, and between then and the late 1990s, had introduced over 300 toys and other products.

Mr. Bev W. Taylor. While a student at Washington University in St. Louis, Bev W. Taylor marketed supplies for magicians. He ran a store, sold by mail, and performed at magic shows. Then, he sold the business and began working as plant manager at Steven Mfg. After Roscoe Zemelman died in 1956, Mr. Taylor bought the company, and in 1964, moved it to Hermann, Missouri, which he had first visited on vacation with his family. In Hermann, Steven Mfg. occupied a former shoe factory building that dated to about 1900. Bev Taylor was a magician before he was an engineer and toy manufacturer, and in the 1990s, he was still performing an occasional magic show. His style of magic, intended to "amuse and amaze" is described in *Bev Taylor's Town House Magic*, a 1993 book by Bruce Hetzler, Ph.D.

Hermann, Missouri. The city of Hermann, Missouri (population: 2,754) was built by German settlers in 1836. Known now for its wineries and German heritage museums, Hermann is a picturesque village on the Missouri River in central Missouri, about 65 miles west of St. Louis. The river, however, contributed to Steven's undoing.

Horses Return in 1983. Steven Mfg. began testing the Hartland horse molds in fall of 1982. In 1983, Steven sent samples of horses to Roger Williams, retired sculptor of Hartland Plastics, for his approval. Mr. and Mrs. Williams told me in 1990 that they were pleased to be included in the process. Steven employee Paola Groeber, the only "horse person" at the company then—she had raised horses as a teenager—was asked for advice, also. Both actions were typical of the conscientious and courteous way Mr. Taylor did business.

Steven began to produce the three 11" horse shapes in 1983, in styrene plastic. Steven photographed some 1960s, 9" series models owned by collector Jamie Glisch, and in December 1983, ordered two copies of my book to help Steven's product development manager, Charles Wuertemburg, identify the horse molds, see the 1960s-1970s colors, and plan the line. At that time, Mr. Taylor told me that Steven possessed all of the Hartland horse molds.

By spring of 1984, the 11" series models were available from several model horse distributors. In March 1994, Mr. Wuertemberg sent me the Cochise/Longley, Semi-Rearing pinto (from the rider series) so I could paint the correct, white marking on the head since Steven was unable to find the paint mask for the model's head. That 9" series horse and some others returned in 1984. In 1985, Steven brought back the 12-piece, Farm Animal Play Set, including the farm horse and donkey, and added a new, 11" series mold, a grazing mare sculpted in China. Except for its grazing pose, it closely resembled the Breyer Yearling in stance and conformation, but Steven didn't know that.

In 1985, Paola Groeber began her small business, Hartland Collectables, Inc. ("HCI"), as a distributorship for Steven's Hartland models. She was then wearing two hats: Steven employee (her title was "sales administrator") and Steven distributor. In 1985 and 1986, it was a bit confusing because she used her maiden name, Paola Skelton, in correspondence between HCI and collectors, but used her married name, Paola Groeber, in correspondence between Steven and collectors and in articles she wrote for model horse journals.

Paola Groeber's article in the July/August 1985 *Model Horse Gazette,* was written from the standpoint of a Steven marketing employee involved in developing the Hartland line. In it, she wrote that from May 15, 1985 on, HCI was having Steven apply clear sealer to every horse model HCI purchased. So, collectors who bought Steven models from HCI (Paola) received models with a semi-gloss finish; purchased from other Steven distributors, those same Steven models had a matte finish. Later, (in the July/August 1987 *Model Horse Gazette)*, Paola Groeber, who had by then consolidated her identity under one name, described the semi-gloss finish as a compromise:

> All our models will have a semi-gloss finish which helps protect the paint finish and seal it from scratches and rubs. Some people liked the matte finish and others liked the high-gloss shine. We feel with a semi-gloss it should make almost everyone happy as they are easy to photograph and [it] still protects the paint finish.

Glossy models can be difficult to photograph, and with the photo shows being a big part of the model horse hobby, it was an important consideration.

Late in 1985, collector Jan Kreischer, eager to encourage Steven Mfg., sent a copy of the book *The Color of Horses* by Ben K. Green to Paola Groeber, who in 1986, was put in charge of the colors for Steven horses. Model color names such as "raven black" and "standard chestnut" came from Green's book.

In 1986, Steven was testing sample colors for more models to be available in Tenite in summer and fall of 1987. The horse line was just hitting its stride when misfortune struck in the form of the worst flood in Hermann, Missouri, in 144 years.

The 1986 Flood. On Saturday, October 4, 1986, for the second time in two months, heavy rains caused floods throughout much of the Midwest. It was bad news for the Steven plant, located along Frene Creek off the Missouri River, in downtown Hermann. A message to Steven customers in November 1986 from Monte Carder, then national sales manager, read:

> On the morning of October 4, our employees and Hermann neighbors lost a two-day battle as the Missouri River crested at 36 feet over is banks. The unforeseen happened. When we had the rising water licked with walls and sand bags, suddenly cement floors ruptured under the fantastic pressure. In minutes, our offices and headquarters were underwater. By a miracle, no one was injured. Two warehouses and our molding facility were saved...The extent of the loss is well into seven figures."

Paola reported that the first floor was under 7.5' of water. Employees (including herself) saved some of the models before the unsanitary, flood waters touched them, but hundreds of finished and unfinished models and much packaging had to be thrown away. After 41 years in the toy industry, Steven wasn't about to give up most of its toy lines, but it did discontinue the horses.

Paola' Christmas 1986 Sale. At Christmas 1986, Paola's Hartland distributorship, HCI, offered a few hundred test color and sample models for sale. She had painted them herself in her role as a Steven employee helping develop the horse line, and then purchased them from Steven after the flood. Included were four horse families: Arabians, Tennessee Walkers, Morgans, and Quarter Horses; it was the first time those models had been made in over a decade. Some models were styrene while others were Tenite, painted in acrylics. They had high-gloss or semi-gloss finishes.

Paola's May 1987 Sale. Also after the flood, Paola bought Steven's entire stock of unpainted models that had been stored in a building that was not touched by the flood. She took them home and painted them as test colors for herself, not as part of her job at Steven. In May of 1987, she sold them to collectors by mail. Those were styrene models.

Production Resumes, and Paola Becomes a Manufacturer. Paola Groeber said that, at her urging, Mr. Taylor resumed horse production in mid-1987, and she announced that she had officially become a manufacturer of Hartland horses, not just a distributor. In an arrangement between her and Steven, she received a steady supply of molded bodies to paint and sell to the small circle of sophisticated, discerning collectors who subscribed to model horse journals and/or were in touch with mail-order sellers of model horses.

Paola painted her models at her home in New Haven, Missouri, about 20 miles from the Steven plant in Hermann. (Paola has moved since then.) Her husband, Larry Groeber, helped her paint models almost every day while he was laid off for three months. Paola's sister, Norma Reed, glued model parts together, cleaned seams, and mixed some of the paint.

Steven was also molding horses for itself again, for the general toy market, and in parts of 1987 and 1988, Steven horses were being sold in Wal-Mart Stores, Inc. In 1987, Paola trained Steven painters in air brushing so Steven "could work to supply wholesale distributors and the toy market" ... while she continued to "serve the true collector who wanted a 'quality' model."

"In an attempt to kick start the model horse program again in Tenite plastic, I gave them permission to produce Steven Hartlands in some of the same colors I was creating for serious collectors." Wearing her HCI manufacturer's hat, Paola invented the colors, and Steven followed suit. She produced samples, documented the paint mixes and procedures, and then trained the Steven painters. She said she did not paint any of the Steven models, however. As a result, there are 16 models by Steven and Paola in identical or near-identical colors, all in Tenite plastic.

Paola wrote in the *Model Horse Gazette*, July/August 1987: "Many of you know the struggle I have had keeping the models alive for the last two years in the Steven toy line. The struggle has now taken on a different ...form...," that of production schedules for models having to compete with production schedules for "hot toys."

(For some unfathomable reason, much of the world thinks other things are more important than model horses! However, a recent—2000—newspaper article listed "collectible horses" among the top seven items in the "girls' aisle" at toy stores.)

Identical Models — and Steven and Paola Distributors.

In identifying models, it is important to note that Steven and Paola made 16 models in identical or near-identical colors, in Tenite plastic, during 1987-1990. Any given example of a model is more likely to be by Steven because the Steven models were produced in larger quantities. In 2000,

Russell Seifert, who was Steven plant manager from about 1989-January 1994, said that thousands of horses were made. About 300-400 would be painted in a day, and that could go on for a month or two, and then the company would produce other products for a month or two, and go back to horses again.

Knowing the source can help identify the model as the painting work of Steven or Paola. Models purchased in stores were by Steven, but models by both Steven and Paola were sold by mail in the 1980s and 1990s. The mail-order sellers of Steven models included Black Horse Ranch, VaLes, The Silver Leash, and Mission Supply House.

Paola was the major seller of her own models, but models that Paola made as model horse show specials were sometimes sold by mail by the show hosts after the show. Paola also produced specials for Black Horse Ranch (sold only by BHR). Cascade Models sold Paola's horses by mail order in 1990 and 1991 until the supply ran out.

After 1991, Cascade Models sold Hartland horses by Steven. Show special models by Steven in the 1990s were also sometimes sold by mail by show holders, and in the 1990s, Black Horse Ranch and numerous other model horse, mail-order outlets sold regular Steven models. The horses in the 1991, 1992, and 1993 JCPenney Christmas catalogs and in the 1994 Enchanted Doll House catalog were by Steven.

Horses in regular run colors made by both Steven and Paola also appeared in the 1989 and 1990 Your Horse Source catalogs. The regular run colors had the Steven item numbers, so apparently, Steven was the manufacturer. Four special run colors by Paola were included, with numbers like Paola's. In 2000, Paola thought that Your Horse Source had gotten the special run models from her (for resale). The specials were Polo Ponies in palomino and chestnut, and Mustangs in black and dapple grey.

Paola's article in the July/August 1987 *Model Horse Gazette* also corrected the list of models sold at Christmas 1986. In many instances, the quantities of models actually sold turned out to be slightly larger or smaller than the quantities published on the sales list. That was partly because they weren't all painted at the time the list was published, so there was some leeway to paint the colors to accommodate the orders although the number of models of each shape was fixed. This book uses the corrected quantities.

Lady Jewel and Jade. In 1987, Paola offered model saddlery by Kathleen Bond, and in 1988, added two new molds, an Arab mare and foal sculpted by Kathleen Moody, the Lady Jewel and Jade molds. Paola said she convinced Mr. Taylor to have Steven commission them. Steven had the molds made in Hong Kong, but the models were produced at Steven's molding division in Missouri, Blanke Plastics. Paola began painting Lady Jewel and Jade and selling them through her catalog in 1988. There were no Jewel or Jade models painted by Steven until 1992. (Steven's molding division was, of course, molding these models for Paola.)

Resin Horses. Independently of Steven Mfg., Paola added resin horses to her line in 1989. She commissioned sculptors Kathleen Moody, Linda Lima, and Carol Gasper to create clay horses which the artists took to a casting service near them, DaBar Enterprises. Paola then painted and sold six styles (shapes) of resin horses that originated in 1989 and 1990.

Paola Ends Business. In 1989 and 1990, Paola was busy attending model horse shows, where she sold models, took orders, and sometimes donated one-of-a-kind models as prizes. Several shows ordered special run colors from her. Paola's arrangement of purchasing molded bodies from Steven ended when Steven stopped production early in 1990. She then left Steven for other employment, also. She continued to paint her existing supply of models, and officially closed her business December 31, 1990.

A model show in St. Louis in July 1990 was the last one she attended, HCI's farewell. Each participant got a souvenir model she had painted, and most were unique or very rare silver, gold, copper, or blue models. She painted all of them on two Saturdays.

Paola said she had never paid herself a salary, and the strain of working full-time and running a business on the side had left little time for family life. She frequently had fallen behind on filling orders. Collectors sometimes had to wait six months or longer to receive models they had paid for. In 1990, Cascade Models, owned by Daphne Macpherson of Edmonds, Washington, started in business as a model horse distributor, and sold models by Paola, including resins. In 1991 Paola filled the backlog of HCI orders. The final models Paola painted that were not pre-ordered were available only through Cascade Models, not directly from Paola. Later, Cascade sold models by Steven.

Stevens in Stores. In 1989 and 1990, Hartlands by Steven were found in many stores with toy departments. There were Tenite models and also styrene models that had been produced in 1983-1986. In 1990, the 9" series models were packed in a clamshell package with a collector card, and the models were named. Paola said that she named the models and wrote the short breed stories that appeared on the collector cards. The idea was to make the models more appealing to children, she said.

Stevens for JCPenney. In spring of 1990, horse production in Missouri had ended because six of the 9" series molds were sent to China, so a factory there could make Steven horses for the 1991 and 1992 JCPenney Christmas catalogs. It was the first time Hartland horses appeared in a major retail catalog since the 1960s. For the 1991 Three-Stallion sets, the Five-Gaiter is styrene, but the Mustang and Arabian Stallion were sometimes styrene, sometimes Tenite. The Tenite models had been 1987-1990 regular runs; Steven evidently supplied those to JCPenney, and had the balance of models made in China in styrene. The Five-Gaiter had to be newly made in order to qualify as a stallion. The original mold was really a mare, so it wouldn't have worked to use up the regular run model (in red roan color) they already had produced. The 1992 Three-Mare sets all seem to be Tenite (in a matte finish), so the existing supply may have sufficed. Perhaps, none of those were made in styrene in China after all.

Mr. Taylor Retires. The molds returned from China to Missouri early in 1992, and in March 1992, Mr. Taylor sold the company to Tim Ford and retired. Tim Ford had worked as general manager of Steven while Mr. Taylor was owner. (In September 1990, Mr. Taylor wrote that the nice comments I'd gotten from Hartland collectors about my book made him "wish we [Steven] could have gotten more involved in horses.") Ford, as president, headed a group of investor-owners. Sam Stone, auditor-comptroller, came from Arkansas when asked to be Ford's partner in the company.

Steven Under New Owners. During 1992-1994, Steven produced (molded and painted) Weanling Foals, two 7" families, and the Lady Jewel and Jade molds. While Steven had previously molded these shapes for Paola, this was the first time it used them itself. Steven also continued to use the 9" and 11" series molds.

The new owners of Steven did not want competition, and Paola Groeber complied with their request to sell her remaining stock of unpainted models back to them. (In 2000, she said she wished she still had them to paint.) They also wanted her to give up her business name, Hartland Collectables, Inc., because they wanted to use it. Paola didn't give it up, but Steven then used variations of it (without the "Inc.") on their catalogs. Despite the similarity of the name, the 1993 and 1994 dark blue Hartland catalogs from Steven involve models with which Paola Groeber had no connection.

Paola's Brochures. Paola's brochures (1985-1990) are always labeled, "Hartland Collectables, Inc.," but in 1985-1987, her sales literature included Steven models because she was a seller of Steven models then. The HCI photo-sheet with the #270 Indian set on it depicts all Steven-painted models. In 1987, Paola briefly wore three hats: Steven employee, seller of Steven models, and manufacturer-seller of her own models.

In 1992, there was no regular line of horses, but there were special runs and the 1992 JCPenney catalog horses. In addition to the Three-Mare sets, there was Roy Rogers on Trigger, and a Lady Jewel and Jade set in bay. These last two sets were molded in Missouri, in Tenite plastic.

Some of the staff was new, and the seams on the Missouri-made, special run horses were not perfectly smooth or had small gaps. Since the 1992 JCPenney Jewel and Jade models had been molded in brown plastic, the seams could not be sanded without the friction turning the plastic white. Gaping seams could not be filled because the filler would show up against the plastic. The flaws were minor, but loomed large for collectors and dealers who had been spoiled by Paola's models.

In 1993, Steven called me and other collectors for advice on what breeds the rider-series horses might be reinvented as. Daphne Macpherson critiqued the sample models for the 1993 line. A new problem—sticky or peeling paint—arose, but the real misfortune was the flood of 1993. The flood came at a bad time because summer is when stores place their orders for Christmas.

The Flood of 1993. On Tuesday night, July 6, 1983, 6-10" of rain fell in Hermann in four hours, and the Missouri River rose to 35.7'. Water covered the Market Street bridge, and it didn't re-open for four days. At the Steven plant, 224 E. Fourth St., sand bags were piled along the retaining wall separating the 100-year-old building from Frene Creek, but at noon on July 8, the retaining wall broke, and the force of the water collapsed part of the wall of the building. The *Hermann Advertiser-Courier* reported that within two-and-one-half minutes, flood water filled the building. The pressure knocked out the windows; and boxes and plastic items were swept outside. Interior walls were knocked down, and part of the front office along Fourth Street was torn out. Interior walls in the research and development building were also knocked out. The newspaper reported that the computers and phones, and much of the portable equipment and raw materials had been rescued the day before, but certain raw materials and equipment too heavy to move were a loss. Steven Mfg. and Hermann, Missouri, weren't alone in their plight: Forty-nine counties in Missouri were declared disaster areas.

In 2000, Mark Borzillo, who had been the Steven mold engineer from April 1993-December 1995, recalled that "no molds were left in the building" at the time of the 1993 flood. No molds were damaged, but at the 1993 Model horse Collector's Jamboree in California, Steven representative Robin Hileman said that the Hartland horse sample models were lost in the flood. Cascade Models owner Daphne Macpherson said that she had been advising Steven on proposed woodcut models in new colors not seen in the 1960s-1970s.

For a while after the flood, Steven was divided between four locations. The offices and research and development department were temporarily housed on the second and third floors at the City Hall Offices. A manufacturer of metal cabinets, House of Metal Enclosures, Inc., made room for Steven's Hartland department in the former Bevco building on Jefferson Street. As Carolyn Laboube of H.O.M.E., Inc. said, "They were displaced by the flood." Steven's Hartland department rented the Bevco space for a year.

There, 20 employees headed by Mark Borzillo made Hartland models along with kaleidoscopes, gyroscopes, and pick-up-sticks. The Steven warehouse in the Big Spring area became the main factory for a while with 80 employees. Wiring and plumbing had to be installed, and machines from the old factory had to be moved across the river. The newspaper reported that for a while, some employees had to cross the river by boat in order to get to work. The fourth location was part of the main building that had been flooded, but could still be used.

New Steven Plant. In fall of 1993, construction began on a new building to house Steven in Hermann's industrial park, high land no flood could reach. It was built jointly by the city, local industrial authority, local banks, and Steven as a measure to keep jobs in Hermann. In September 1993, Steven had 122 employees, and expected to reach 150 by the middle of 1994.

Production Resumes. When Hartland production resumed after the flood, a new problem developed. Some of the models issued in 1993-1994 suffered from incompatibilities of paint and plastic or paint and gloss coat, according to Mark Borzillo, Steven plant engineer. The result was paint that peeled or flaked off after several months, was sticky, or melted in 95 degree heat. Because of the disruption caused by the flood, it took the company a while to realize that there was a problem.

At first, they thought the problem was that, due to the shortage of storage space after the flood, models were being shipped before the paint had a chance to cure. Then, there was the problem of rubs caused by the plastic bands that were needed to hold the models in the new style of box. This problem was seen with the palomino Tennessee Walker Mare and Foal set for the 1993 JCPenney catalog. Steven corrected the problem by placing soft wrap between the bands and the model and between the model and the inside of the box. The paint-gloss-plastic incompatibilities were actually a separate problem, and one that was not obvious since it might not show up until months after the model was shipped. In an attempt to economize, Steven had mixed and matched brands of paint, gloss, and plastic, and some combinations of them were not compatible, as it turned out.

Under national sales manager Don Light, Steven offered prompt and courteous refunds or exchanges in 1993 and 1994, but there were unavoidable delays later. Not just the Hartland department, but the entire company had been financially mismanaged during the time Tim Ford was owner and Ken Movold was in charge of the Hartland line. After Tim Ford quit in fall of 1994, the staff was loathe to mention his name.

By October 1994, Steven was consolidated in the new building at 104 Industrial Drive in Hermann with its 44,000 square feet of warehouse and shipping department, 39,000 square feet of air-conditioned production plant, and 6,000 square feet of offices. The old plant and research building were torn down and replaced with a soccer field.

Production Ends, Leaving Models Unmade. The final Hartland models were molded at Steven until August or September of 1994, and the last of them were painted in November 1994.

To the disappointment of collectors, some special runs were never issued. Only one sample model was made of the liver chestnut Polo Pony that was to have been a 1994 special for an East Coast riding club affiliated with collector Patricia Henry. Only two samples were made of the silver metallic Mustang that would have been a special of 2,500 for Cascade Models for the 1994 Northwest Congress model show. The 1994 Jamboree model was supposed to be the 9" Arabian Stallion in dapple rose grey, but those models were never made, and the "bay roan" Weanling Foals were substituted.

Although they were depicted in the 1993 and 1994 catalogs, the Friesian and Mule sculpted by Kathleen Moody were not issued, and the molds for them were not finished. In addition, five more clay originals that Steven owned were not molded. They were a Percheron by Linda Lima, and a Trakehner, Morab, Missouri Fox Trotter, and leaping Lipizzan by Kathleen Moody.

Mr. Taylor Buys Steven Back. In March of 1995, Bev Taylor came out of retirement to buy back Steven Mfg. (His wife, Frances, had died, and he wanted to keep busy.) He weighed the possibilities of producing Hartland horses and sports statues again, but also entertained offers to sell the Hartland line. With the financial setback of the flood, Hartland production would at least have to be postponed.

In 1995, Steven almost had a buyer for Hartland. As I wrote in *Hartland Market*, December 1995:

December 14, 1995—There's good news and bad news at Steven Manufacturing Co. Like an almost certain home run that goes foul at the last second, or a [winning] horse that collapses just before the finish line, the sale of Steven's Hartland division fell through on November 30 after five months of negotiations. The only impediment was that the buyer was, finally, unable to come up with the money.

The mood was somber at Steven Mfg. Although the company is, as one manager [Mark Borzillo, engineer] expressed it, "emotionally attached" to Hartlands, management seemed as disappointed (and surprised) as many collectors probably are at this news. The sale would have included the Hartland name, rights, molds, models in stock, and the seven clay equines for which molds have not yet been made.

The good news is that [Steven] is now seriously considering producing horses, rider sets, and possibly [the sports] statues again, starting in [about two years].

Steven owner and president Bev Taylor hinted that the baseball part of Hartland might be sold off separately after all. When asked about the seven clay equines, he said they'd be likely to keep them along with the rest of the horse line, but he asked how much I thought they were worth.

I replied that Kathleen Moody was paid $650 (actually $620) for each of her six, but "they're worth more than that" (assuming they're still in good condition).

He said, "$1,000?"

I added that Kathleen had said she wished she could get the models back so she could improve them.

Mr. Taylor laughed and said. "I wish we could get this settled."

He could have been referring to the entire business of determining the fate of the Hartland part(s) of his company.

In a year and one-half, Steven had entertained eight groups of prospective buyers for Hartland, but only one came close.

In 1995, the clay horses were believed to still be "in a drawer," but early in 1996, Steven officials reported that the clay equines had apparently disappeared. Tina Strubberg, who was chief painter of Hartlands and one of the last Hartland department employees when she left in June 1995, said that she last saw them after the flood of 1993, but before the move to the new building in fall of 1994. Liz Dalpe, national sale manager, said that no one knew what had happened to them. They were not returned to the artists. In 1993-1994, a lot of portable Hartland property, such as file photos, apparently disappeared with departing employees.

In April 1996, Kim Lorraine, executive assistant, reported that Steven officials had gone to Hong Kong earlier in 1996 to consider having Hartlands made abroad. In May 1996, Sam Scott, chief financial officer, said that there was, however, "nothing new on the horizon" for Hartland.

New Owner in 1997. Finally, at the end of February, 1997, Mr. Taylor sold Steven (including Hartland) to David Segal, and retired for the second time. Since then, Mr. Segal, who also owns at least one other toy company, has not been interested in Hartland. According to local observers, by 2000, Steven has shrunk to fewer than 20 employees. The new building at 104 Industrial Drive, which a lease-purchase agreement permits Steven to own after 15 years, is now almost devoid of manufacturing activity. After the long history of Steven, it seems a shame. *Shortly before this book went to press, Hartland (but not Steven) was sold to a new owner, who plans to revive the horses and other Hartland models. Please check my web site--www.execpc.com/~gfitch/Index.htm--for news updates.*

Summary of Hartland Horse & Dog Model Eras:

Hartland Plastics, Inc., Hartland, Wis.
        Horses begin: late 1940s
        Large Dogs begin: 1950s
        Breed series horses and small dogs: 1960s

Durant Plastics Div. of Strombecker Corp., Durant, Oklahoma
        Horses and small dogs, 1970-1973.

Steven Manufacturing Company, Hermann, Missouri
    Owned by Bev Taylor
        Horses in styrene: 1983-1986.
        Horses in Tenite: 1987-1990.
        Horses made abroad for 1991 JCPenney catalog.
    Owned by Tim Ford, et al.
        Horses in Tenite: 1992-1994.

Paola Groeber's Hartland Collectables, Inc.
    Sells Steven models only: 1985-fall 1986.
    Paints and sells HCI models:
        Horses: Christmas 1986 sale
        Horses: May 1987 sale (styrene)
        Horse production: 1987-1990
        Finishes orders: 1991
        Bay Resin Pasos: 1992
        December 1994 sale of models from 1987-1990.
        Fall 1999 sale of 1987-1990 models
        Spring 2000 sale of 1987-1990 models.

Hartland made horses for a variety of tastes: heights from 2" to 10", builds from refined to robust, horses in action and in repose, numerous breeds, and colors from realistic to imaginative, including wood-simulated horses.

The Hartland Magic

Hartlands were originally marketed as toys—that was Paul Champion's idea—but were rediscovered as collectibles when other so-called toys were in the 1980s and 1990s. Model horse lovers always appreciated them for their beautiful representation of horses: the artistry of the sculpture, the beauty of the colors, and the durability and affordable price of their medium—plastic.

Hartland Plastics' Robert McGuire was primarily responsible for the styling of the horses; that is, he decided upon the shapes—breeds or poses—to be made and their colors. Hartland Plastics' two sculptors, Roger Williams and Alvar Bäckstrand, created individual horses and horse families with recognizable breed characteristics and accurate and flowing gaits. There were no awkward, cross-cantering models or "mystery gaits" unknown to science. They also sculptured Hartland's dogs and all other Hartland models that originated in the 1940s through the 1960s. Roger Williams was especially known for his ability to capture the expressions of animals. The detail on Hartlands is outstanding: note the muscling on the shoulders of the 11" series Quarter Horse. Hartland's unusual mold-making process allowed fine detail to be built into the original art and then reproduced successfully.

Hartland's breed-series horses are easy to recognize as members of their breeds. There are Arabians, Thoroughbreds, Saddlebreds, Quarter Horses and Appaloosas, Tennessee Walkers, Morgans and Pintos, Mustangs, donkeys, a Belgian draft horse, and a polo pony. There were no true ponies, but gaited-breed lovers could hardly have asked for more.

The poses, from the fighting Mustang to the romping foals, are imaginative, and Roger Williams also had a flair for depicting the locomotion of animals. The elastic trot of the 9" Three-Gaiter and the fluid, running walk of the 7" and 9" Tennessee Walkers are typical of the graceful poses and accurate equine gaits.

The colors and finishes of Hartland horses have varied, depending on the manufacturer at the time. The original company set a precedent for colors so beautiful that horse lovers wouldn't care whether they were realistic or note. Hartland Plastics applied "pearl white paint" to some 9" and 11" horses in the 1960s, and Paola Groeber used paint with metal flecks to create "pearled" colors among her 1986-1990 models. Both are now among the most prized model colors. Some 1960s horses had high-gloss finishes while others were satin (semi-gloss) or matte. Satin and matte finishes are most common on 1980s-1990s Hartlands.

Few of the colors of the 1960s models are entirely realistic—real horses aren't glossy, bright red, blue, pearl white, or metallic copper. They don't look carved from wood, nor molded in solid yellow—but that is part of their charm. Some of the colors improved on nature, such as the sorrel, Arabian stallions with the lovely, but unrealistic, flaxen-colored stockings to match their accurate, flaxen mane and tail. If too much realism could be boring, Hartland horses did not present that danger. Hartland used artistic license to create collector models for horse lovers to dream on.

The original horse designs from the 1960s have been used through the 1990s, but two, noteworthy horse shapes by sculptor Kathleen Moody were added in the late 1980s. Resin horses in new breeds were also added to the "Hartland" line, and 9-11" horses were sold in unpainted colors—for collectors to paint themselves—for the first time. Some very complicated and realistic horse colors were produced in the 1980s and 1990s. The result is a delightful collection of models for horse lovers to admire.

New Artists and Models in the 1980s-1990s

The Resin Horses. In the late 1980s, Paola Groeber's Hartland Collectables, Inc., helped usher in the "resin revolution" in the model horse industry. Most of the model horses available up to that point had been mass-produced in plastic or china. Only a few artists were producing limited edition, resin-cast horses when Paola added them to her 1989 and 1990 catalogs. Paola more or less "discovered" Carol Gasper, Linda Lima, and Kathleen Moody. Hartland Collectables, Inc., added the six resin models because the injection molding process (with steel molds) "would not have been able to duplicate their many undercuts and details."

Carol Gasper and Linda Lima. Carol Gasper sculpted the "Pocket Ponies" which Paola Groeber then painted: a show-type Tennessee Walker at the running walk and a standing Quarter Horse with its mane blowing in the wind. Linda Lima, Carol's twin sister, sculpted the walking Arabian Gelding that Paola painted and sold in dapple grey.

Linda Lima, a graphic artist, started drawing and painting horses professionally, including in oil and pastels, in 1975. From 1984-1989, she also painted replicas of antique carousel horses that her husband, Sam Lima, and his son hand-crafted in fiberglass; the replicas were cast from the antique, wooden originals. After Sam died, she continued to paint and sell the replicas until 1995.

Carol Gasper said that both she and her sister, Linda Lima, began equine sculpture about 1986. They have been drawing or painting horses since grade school, and each majored in art. Both have collected wooden carousel horses, and both owned real horses for at least 20 years. They live near each other and Kathleen Moody. In 1990, Carol said, "The three of us like to get together to share sculpting ideas and see each other's new figures."

Kathleen Moody. Kathleen Moody, another lifelong horse lover and equine artist, collected model horses from about age 6, and was modeling horses out of clay in junior high. Between then and 1982, she did wildlife taxidermy for about 17 years, which she said gave her a firm foundation in animal anatomy. Also, before she married, she worked for a division of Walt Disney studios, doing sculpture, costume creation, painting, and model miniatures work. Since then, she's been a freelance equine artist and has become the mother of eight. For a time, she boarded other people's horses to save money to become a horse owner herself, and then enjoyed entering horse shows. As her family grew, she cut back on real horse activities in favor of model horse collecting, remaking, and showing, and the stage was set for her emergence as an equine sculptor for the collector market. Kathleen's equine resin sculptures for Hartland Collectables are the Miniature Horse, Peruvian Paso, and Simply Splendid, the cantering Arabian Stallion.

Resin is Durable. Resin is durable, and Kathleen Moody called it "unmatched for capturing intricate detail." In addition, start-up costs to make a resin model are cheaper than to mass-produce a plastic model by injection molding. No factory is needed, and no heating or melting is involved; it can be done in a garage or kitchen. A disadvantage is that the quantities that can be made are limited, and the cost per unit to produce a resin horse is higher. The process involves a silicon (rubber) mold, which typically breaks down after about 50 pourings. Resin is a pourable plastic similar to fiberglass, but without the fiber, Kathleen said. The materials, silicon and resin, are expensive, and a skilled craftsman is required. For an amateur at moldmaking, the horse figures are complicated. They "require a lot of seam cleaning, repair of badly molded parts, and detailing," Kathleen said. Thus, resin horses often are priced at $150 or more whereas a plastic horse the same size might cost only $15.

Injection-molded, plastic horses cost much less per unit because thousands of them can be made after the huge, start-up costs are overcome. Kathleen said in 1990 that the cost of a plastic injection mold (in steel) was $15-20,000 or more, plus the cost of the hanging frame to fix the mold into the injection molder, highly-skilled labor, machine maintenance, and the cost of Tenite plastic. It would be more now. (To start up a new molding factory would probably require a large group of investors, and advice from consultants already in the industry.)

Paola Groeber had the artist's models cast by a resin-casting service, DaBar Enterprises, which although successful, closed in the late 1990s. All three sculptors issued new designs of resin horses after Paola closed Hartland Collectables at the end of 1990. Since then, more than 50 additional sculptors have created over 200 horse designs that have been cast in resin and sold as limited editions. The resin horses are recorded in a book, *Resin Registry*, by Julie and Frederick Harris, 1997. The book is loose-leaf to accommodate annual updates.

Lady Jewel and Jade. In 1988, Paola Groeber added two, beautiful new models to the Hartland line: a cantering, Arabian mare and running foal by Kathleen Moody. Acting as Kathleen's agent, Paola convinced Steven Mfg. to commission the Lady Jewel and Jade models and mass-produce them in plastic. The molds were made overseas (in China), but the models were molded by Steven in Missouri.

Jewel and Jade were time-consuming to produce. Paola wrote in her September 1989 flyer, "We have been putting them together by hand, and painting the manes and tails by hand as we do not have gluing fixtures or paint masks to help us." The Jewel and Jade models were signed on the belly with the Hartland Collectables logo, a heart with "HC" in it, and the month and year they were painted. There were no Lady Jewel or Jade models painted by Steven until 1992.

The third new horse mold Steven commissioned in the 1980s was the 11" series Grazing Mare. The sculpture and mold-making were done in China, but the models were molded in Missouri.

New Colors. There were no dapple grey Hartland horses until the 1980s. The first dapple greys were painted by Paola in 1987, and then Steven painted dapple greys, too. Paola Groeber also made some dappled palominos. (In nature, a healthy horse of almost any color can show dappling. Unhealthy horses never do, but many healthy horses are not dappled, either.) Fine, horse color details—such as dorsal stripes and shoulder and leg stripes on dun horses, and the dark heads on roan horses—were rarely attempted by model horse manufacturers until Paola and Steven Manufacturing did them. Paola painted these details freehand, with excellent precision; the Steven details on the same model colors were sometimes a little less exact, less subtle, and more variable from model to model, but you have to give them credit for the extra effort.

Seven More Equines Not Produced. In 1988, sculptor Linda Lima sold a Percheron she had sculpted to Steven Mfg., but it was not produced as planned. She said she had forgotten whether it was a stallion or a gelding. She said that any similarity to Breyer's "Roy," is a coincidence since "Roy" came out in 1990.

Likewise, Kathleen Moody sculpted a Mule, Friesian, and four other equine models that Steven did not produce before ending Hartland production in 1994. The Friesian, modeled after John Meliot's stallion, Sander, was done in the 11" size, and Kathleen said she thought that Steven had it "shot down to the 9" size on a pantograph (an industrial copier, reducer, and enlarger for 3-D objects)" ... after they "changed their mind and wanted the 9" size, instead. She then sculpted the other five equines in the 9" size. None of the seven models was ever produced although the Mule and Friesian did make it to the catalog.

That was in 1989-1990 and Paola Groeber, who then worked at Steven (in addition to producing her own line of Hartland models), selected the breeds. "It was to be the 'Hartland Gems" series, with Lady Jewel and Jade being the first," Kathleen said. Later, Paola made the connection between Kathleen Moody and Breyer, which has produced several models by Moody.

Kathleen Moody credited "Sheryl Leisure and her friends" for encouraging Steven to mold the seven clay horses. "Without Sheryl, the Mule and the Friesian wouldn't have made it as far as they did," she said. (Sheryl Leisure published *The Model Horse Trader* for 10 years, and she still hosts the annual West Coast Model Horse Collector's Jamboree. In addition, Reeves International, parent company of Breyer, now employs her to put on the annual BreyerFest in Lexington, Kentucky.)

Since the 1990s, Carol Gasper, Linda Lima, and Kathleen Moody have issued many additional horse designs in resin, amidst the surge of artist resins.

The Original Hartland Artistry

Except for the handful of recent molds (three horses and 11 baseball players), all Hartlands were the work of two, full-time sculptors: Roger Williams and Alvar (Al) Bäckstrand. Roger Williams, a noted sculptor and woodcarver, worked for Hartland Plastics from 1944 or 1945 until retiring in 1968. Alvar Bäckstrand was an accomplished woodcarver, furniture designer, author, and illustrator in Sweden before moving his family to the U.S. in 1950. He painted murals and worked at Haeger Pottery in Dundee, Illinois, before coming to Hartland Plastics in 1956. After model production ended in 1969, he stayed on to design cabinets, other functional products, and art for point-of-purchase displays until 1977.

Hartland's executives (Paul Champion and Robert McGuire) selected color schemes for the figurines, but all art work was done by Roger Williams and Alvar Bäckstrand; they designed and sculptured the Hartland statuary. Both of them worked from photos they took at horse farms, at Milwaukee County stadium (of baseball players), and from photos sent by TV studios. They also watched TV to see multiple views of the actors.

Roger Williams

Roger Williams sculpted most of Hartland's animals, and more than half of its riders, ballplayers, and other figurines. When I reached him by phone in fall of 1990, he and his wife, Idella, had a difficult time recalling specific models because he had made so many. When he retired in 1968, Roger Williams entrusted his log of models to Alvar Bäckstrand, who kept them and his own sculptor's log for 19 years. Because I didn't connect with Mr. Bäckstrand until 1990, the record of when each model was made from

1956-1968 was lost by a margin of only three years. Fortunately, each sculptor saved some other documentation from Hartland Plastics.

Roger T. Williams—who used his middle initial so his name would not be mistaken for pianist Roger Williams—was born May 11, 1903, in Waukegan, Illinois, but grew up mostly in Rhinelander, Wisconsin. From watching his grandfather, a wheel wright, make wagons, tools, and cabinetry, he aspired to someday shape and mold things with his hands, Mrs. Williams wrote. However, "Roger did not pursue a career in art as a young man." His first job was selling insurance in Chicago.

He attended classes at the Art Institute of Chicago for six months, then quit and worked in clay on his own, modeling the human form and animals. Inspired by popular illustrative painters of the day, Williams produced character studies in plaster that were sold in stores on consignment. However, sales were slow during the Depression. For about 16 years, Roger Williams earned a living as a timekeeper at the Chicago & Northwestern Railroad station, first in Chicago, and then in Milwaukee.

"It was then, in the 1920s and 1930s, that Roger started a reference file that during the next 50 years was to fill two large filing cabinets," Mrs. Williams wrote. "Any picture (painting or photo) that appealed to him, he saved." After they married, she took over the reference file compilation.

In 1938, Williams left the railroad to join Ornamental Arts & Crafts Co. in Milwaukee. There, he modeled wildlife figurines that were cast in plaster and sold throughout the midwest. He also made models reproduced in compressed paper for Pulp Reproduction Co., Milwaukee.

Roger and Idella (Mueller) Williams met in 1941 when Ornamental Arts and Crafts hired her to trim the plaster casts. They married in 1944, and Williams began his collaboration with Hartland Plastics in 1944 or early 1945. From their home in Hancock, Wis. (near Plainfield), Williams sculpted Hartland's ornamental picture frames and mirror frames and the company's earliest religious figurines before moving closer to Hartland, Wis., in 1948 or 1949. They lived at Pewaukee Lake, near Edward and Iola Walter, owners of Hartland Plastics, until 1958. Hartland paid Williams per model until about 1948 or 1951, when he became a salaried employee, but he always worked at home. Idella Williams assisted in the model-making process, chiefly by putting the fine layer of clay in the plaster casts.

Roger Williams produced a new model for Hartland about every two weeks. Until 1965 he also continued to furnish plaster casts for Ornamental Arts & Crafts, where he was a partner. Ornamental's 1956 catalog of animal plaques (including horse heads), lamps, and figurines was mostly his work, Mrs. Williams said. Williams' non-Hartland work was usually reproduced in plaster, and he did not make models that were copies of, or would compete with, Hartland models. Mrs. Williams said that Breyer Molding Company had approached him at one point, but he declined since he was working for Hartland. He retired in 1968, and Hartland Plastics discontinued its toy and model horse line, but paid him for other art work (such as an ornate, cosmetic container cover) on a per-item basis until February 1970.

In retirement, Roger Williams again had time to create art purely for enjoyment. He painted in oil and pastels, cast clay models in stone powder mixed with water, and worked in wood, marble, and metal. His woodcarvings won awards at National Sculpture Society shows in New York in 1972 and 1973. His 1972 prize winner for creative sculpture was "Stubby," a 20" long draft horse carved in oak; "Giraffes" won the prize for bas relief in 1973. ("Stubby" was purchased by Charles Parks, sculptor of life-sized bronzes, and "Giraffes" was bought by sculptress Harriet Frishmuth, whose marble statues can be seen at Brookgreen Gardens in South Carolina.) Another year, Williams entered a two-foot-high Thoroughbred foal, but it broke enroute and was not exhibited. Williams' remarkable wood sculpture of a pair of horses joined by their harness ("The Logging Team") appeared on the cover of the National Wood Carvers Association's *Chip Chats*, July-August 1981. "Sad Sack," a long-eared hound, was on the cover of *Sculpture Review*, Spring 1986. He was elected to NSS in 1972, and also was a member of the Society of Animal Artists.

In 1982, when Steven Mfg. prepared to revive the Hartland horse line, it invited Roger Williams to sculpt a new horse model. The then 79-year-old sculptor was excited about it, but when his wife suggested that Steven might want an entire series of horses, anxiety overcame his excitement, and he declined. In 1984, Steven Mfg. sent brochures and sample horses which Mr. and Mrs. Williams returned with their approval of the "beautiful painting job" on the samples.

Williams carved until 1984, when he was 81, and painted in pastels until he was 83. In 1988 and 1989, Roger and Idella Williams auctioned about 150 pieces of his wood and stone sculpture: character studies, nudes, and animals (but no horses). The Williams had "lived like hermits" at a wooded, 10-acre hillsite in Ashippun, Wis. (near Neosho), from 1958-1989. Until the auction, which was nationally publicized, the neighbors didn't know that they'd had a sculptor in their midst. Williams never sought publicity, but appreciated recognition.

In his final years, Roger Williams' eyesight was impaired by macular degeneration. He died April 9, 1991, at the age of 87. In 1992, Idella Williams set up a two-part award program in his name with National Sculpture Society. Every year, a scholarship to study sculpture is given in his name, and the Roger T. Williams prize is awarded to "a young sculptor who reaches for excellence in representational [realistic] sculpture."

Thomas Caestecker, son of Hartland owner Charles Caestecker, said Roger Williams was "very modest"—and remarkable. After lightning struck a tree on the Williams property, the artist carved a five-foot-tall, standing bear from its trunk. Likewise, trees felled by wind or ice were recycled into art.

A riding mishap at the age of 17—a horse dragged him two or three miles—badly injured Roger Williams' knee and left him with one leg two inches shorter than the other, but the incident didn't dim his view of horses. The last carving he worked on, but never finished, was a horse.

Alvar Bäckstrand

As reported by several former Hartland personnel, Alvar Bäckstrand had a special talent for sculpting faces that really looked like who they were supposed to be. Janet Gerbenskey, a spray painter at Hartland Plastics from 1955-1975, recalled that Alvar Bäckstrand "was an excellent sculptor. His work was really great."

After sculpturing the original figure in clay, Bäckstrand also cast and finished the metal model for electroplating, a step in the mold-making process. Bäckstrand prepared his own and, from 1956-1968, Roger Williams' sculpture for molding. He sometimes modified Williams' work so that it would mold more easily. An example is that the 9" Five-Gaiter's tail is angled to the left so that it could be molded with the left half of the horse's body.

When asked how long it took to make a model in clay, Alvar Bäckstrand said in 1990, "It took me about a week, average time to make the clay model, and a week to cast the two halves in model metal, polish it up, and sharpen the details. I think it took Roger Williams a little longer, but it's impossible to tell since he worked at home. I did, too, the first years, but then I thought it better to work in the factory since that gave me some spare time at home."

To research horses, Alvar said, "I sometimes went out and took pictures of horses on farms." He made about two or three, possibly four, horses.

In 1990, Alvar Bäckstrand described his career:

Growing up in Vetlanda, Sweden, he began to carve wood figures as a teenager, but that competed with his other interests in writing and illustrating. He took art courses and learned from some friends who were well known artists. He started to carve professionally at age 27 for a wood sculpturing studio before starting his own furniture design and carving studio. The so-called Swedish modern furniture style became popular before 1930 and was still popular when he left Sweden, Alvar Bäckstrand said. He, his wife Henny, a son, and a daughter (they now have three grown children) arrived in Illinois in May 1950.

"I had been corresponding with the Chief Designer, Olson, at Haeger Pottery in Dundee, Illinois, so I had a job waiting for me. I sculptured some figures there, and they were very well satisfied with my work, so they told me, but I didn't care too much for moving to another country and working for a fifth of what I made in Sweden." So, he resigned.

After that, Alvar Bäckstrand worked in machine shops and display studios. For a while, he ran automatic screw machines 12 hours at night, worked four hours per day in a pattern shop, and refinished furniture on Sundays. In his "spare time," he painted murals in churches and homes in Batavia and Wheaton, Illinois.

In 1956, he was introduced to Hartland Plastics by a friend who was related to Robert McGuire. He started working for Hartland in June 1956 and retired in March 1977 at age 65. Haeger Potteries' loss was Hartland Plastics' gain.

At various times, Alvar Bäckstrand wrote poems and stories for local newspapers (in Sweden), illustrated books and magazines, wrote articles and stories for "an annual book and a magazine about old times," and did artwork for book jackets. After retiring from Hartland Plastics, he made some models for Sussex Plastics, Inc., Sussex, Wisconsin, which is 10 miles from Hartland.

(Sussex Plastics was started by Lorand Spyers Duran and Hans Seuthe after they left Hartland Plastics in 1977. Spyers Duran was president of Hartland Plastics from about mid-1972 to 1977, a year before it closed. Hans Seuthe was Hartland's long-time molding foreman.)

Alvar Bäckstrand also was doing two-dimensional art for several years prior to 1995. He and his wife, Henny, showed me some elaborate, ink drawings with foliage and gnomes in them. After that, his interest turned to computers.

Excessively modest and very hard on himself, Alvar Bäckstrand said in 1990 that most art is 1% talent and inspiration and 99% hard work. On what it's like to be a sculptor, he said, "You concentrate on shape and dimensions to your utmost, and there's still something missing....So you ask somebody, 'What do you think?' hoping that somebody else can see that indefinable something that makes a character come alive."

A Combined Effort

Sculptor Alvar Bäckstrand said that he made only a few horses, and Roger Williams made the rest.

It made sense that Roger Williams made [most of] them. He was a truly great sculptor, especially in sculpting animals. As with all other models, the plastic horses do not do him justice. In the first place, you are severely limited in expressing shape and action since the model has to come out of the mold in a straight pull. The executive insistence that the model be polished worse than an army boot, the grinding away of the undercut in the finished copper mold, and finally the death knell for all traces of art by the applying of bright toy colors.

When Roger brought in a clay model for approval, everybody was awed by it, or should've been. The smoothness and sharpness of details was astounding. I heard many times that he had a modeling secret that he would not tell to anybody. That was not true. He just worked with modeling tools, and brushes wet very slightly in kerosene or spit. He laughed when he told me how to do it, and said something about hygiene and that I could use water as well. . . .

It takes many years of practice to get Roger's skill in working with the clay. Only in the last years did I approach his skill somewhat. Approach, not reach. I made up for it by working better in metal. However, he never seemed to understand the importance and necessity of eliminating all undercuts on the models. He would bring in a model and say, 'I got it out of the plaster mold, so it should work fine.' There is quite a difference in wriggling a model out of the plaster mold, or getting the plastic out of a machine mold that moves in a straight line.

In the later years. . . I changed parting lines on the metal models he made, improved details, and eliminated undercuts. . . . On some baseball and football figures of his I cut off the legs, added more metal, and lengthened the legs that were so out of proportion that I couldn't let them go. From 1956 until the sorry end of Hartland Plastics, not a single model left the factory that I had not worked on.

In regard to the alterations, Bäckstrand recalled Williams saying to him, "I trust you completely." When I interviewed them in 1990, each sculptor praised the other's work. (At that time, Mrs. Williams was making ceramic frogs and thimbles in a kiln at their retirement home. She still does volunteer work, makes and sells crafts, and donates the proceeds in her husband's name.)

Roger Williams' name appears on some of the boxes for 9", breed-series horses, but all Hartland models from 1956-1968 were really a joint effort.

Made by A Special Process

Hartland models were injection molded in plastic using electroformed copper molds, rather than the machined steel molds used by most plastics companies (then and now). As a result, details could be reproduced better, and the mold-making process was also faster. Costs were kept down because everything was done under one roof. An article in *Modern Plastics* (October 1960, page 172) said that Hartland Plastics could take an idea through the finished, marketed product in 120 days. It didn't hurt that the paint dried quickly—in 15 to 20 seconds. Precision painting was done with copper, paint-spraying masks to neatly define the painted areas and make them uniform from one model to another. The products were of high quality, and Hartland Plastics was a success story.

Mold Making. For their patient explanation of Hartland Plastics' more or less unique, mold-making process, I'm indebted to Hans Seuthe, Alvar Bäckstrand, Idella and Roger Williams, Edwin Hulbert, Paul Champion, and Richard Rohde. Ten years after they began describing it to me, I think I finally understand it.

Keep in mind that Hartland models are mainly hollow, with a thin layer of plastic. They were usually molded in six parts: one each for the left and right half of the horse, and an extra part for the inside of each leg. However, it is easier to think of them as molded in just two parts: the left and right halves. To picture them, I think of two (kitchen) measuring cups of equal size. When you balance one cup's rim on the rim of the other, with the two hollow sides in, that's like a Hartland horse lying on its side. The "parting line" (seam line) between the two halves of the horse goes along the middle of its back, neck, face, belly, etc.

The making of a model involved three mold-making stages prior to actually molding the horse. (1) The sculptor would create a clay sculpture and, through several steps, convert it to a metal horse in two, hollow halves.

(2) The electroplating department used the metal horse halves to make the inner and outer copper shells (mold surfaces) for each half of the horse. (3) The tool makers took the two copper shells for each horse half and built the finished mold with them. Then, the mold was placed in an injection molding machine, and plastic halves of horses were made. The assembly department then glued together the parts to make a complete horse, ready to be painted.

From Clay Model to Metal Model Halves (Stage 1). Since the clay sculpture was solid, it was necessary to, in essence, take out the center, Richard Rohde said. As sculptor Alvar Bäckstrand said, "The original, master horse model in clay—or in wood as in one case [the 7 1/2" Budweiser Clydesdale]—was made into two or more parts of metal shells by either Roger Williams or me." After the clay model was approved—for Roger Williams' models, Paul Champion or Robert McGuire sometimes went to his home to see the model—the sculptor converted the clay model to a plaster mold, and from that, to a hollow, metal model (in Cerro-based Monel metal) in two halves that, after some adjustments, was given to the plating department.

Steps in Going from a Clay Model to Metal Model Halves. After the clay model was approved, the steps in stage 1 of mold making were these: (a) The sculptor placed the solid clay, original sculpture on its side in a box, and poured plaster of Paris in the box so the horse was buried to the parting line—so high as the middle of its back and belly. (b) After the plaster hardened, he coated it with a parting agent, such as Vaseline. (c) With the clay horse still half-buried in plaster, he poured more plaster over it so the entire horse was buried in the box. (d) Then, he pried apart the two plaster mold halves and removed the clay model, which would be more or less damaged. (e) Taking one plaster mold half, he put a layer of clay in it to the desired thickness of the eventual plastic model's wall, often about one-tenth of an inch (.1"). (f) He put Vaseline over the clay layer, and then poured plaster into the plaster mold, over the top of the clay. (g) Then, he repeated (e) and (f) with the other half of the plaster mold; i.e., the plaster mold for the left side of the horse if he had already done the right side. The result is a "sandwich" with plaster top and bottom and a layer of clay in the middle. (h) The sculptor then took apart the sandwich (mold) and removed the clay. (i) Next, he clamped the top and bottom of the empty sandwich together and poured Monel metal into it. (j) After the metal cooled and hardened, he opened the plaster mold (sandwich with Monel metal in it), and removed the metal horse half as gently as possible. Remember that the metal is only about .1" thick. (k) He repeated (i) and (j) with the plaster mold for the other half of the horse. (l) Some damage would occur to the metal horse halves in this process, and the sculptor would fix them to look as the original clay horse had looked.

Sculptor Roger Williams worked at home, and his wife, Idella, helped with the plaster and metal steps. She said they used a hot plate to heat the Monel metal to 150 degrees for pouring. Sculptor Alvar Bäckstrand performed the same steps in the model-making department, and model-maker Armin Rohde assisted him by pouring the plaster molds, filling clay thickness into the first plaster molds, and casting model metal. (The brand name of the model metal was Monel metal.)

Then, the mold engineers made blueprints for the metal machine mold that would be used in the injection molding machines, Alvar Bäckstrand said. The blueprints included the cavity layout and the cooling system. "We in the model department put the Monel metal shells on plates and pins so they were tilted right and there were no undercuts." Those model halves on plates were then taken to the electroplating department.

Electroplating (Stage 2). Hartland Plastics' molds were called "electroformed" because they were made by electroplating, in which an electric current sent through a chemical bath would ionize copper in the bath, causing it to be drawn to the Monel metal shells and build up on them in a relatively even layer over a period of days. The Monel metal shell for each half of the horse (left and right sides) was copper plated on both sides (the inside and the outside, in turn). The horse halves were copper plated up to .75" thick, Hans Seuthe said. After the Monel metal was pried apart from the copper layer that enclosed it (like unsticking the cheese from the bread in a cheese sandwich), the copper layers ("shells") were used as the molding surfaces in the mold that was then constructed by the tool making department. The electroplating was sometimes done in nickel or a combination of copper and nickel, Edwin Hulbert said. The electroplating ("plating") department, was the key to Hartland's mold-making process, Hulbert said. The mold replicated the contour of the metal model at the molecular level; it's not possible to make a more exact copy than that!

Steps in Electroplating. Alvar Bäckstrand said that, in the plating department, the Monel metal model halves on plates and pins:

"...were put in wax up to the parting line, then sprayed with a conductive silver paint. This wax and metal mold model was then electroplated from one-eighth to one-quarter inch thickness. This made the first male machine mold half. The wax was then removed [melted away] and the copper shell with the metal model in it was then, again, sprayed with silver paint which was a parting agent between the male and female halves of the copper mold, as well as a conductive for the plating."

The copper deposited on the silver lacquer (paint). The electroplating was done in two steps, first the "inside" side of the hollow, metal horse half and then the "outside" side of it. After the electroplating was finished, the metal half-model would be like the center of a sandwich with copper for the bread. The parting agent allowed the center of the "sandwich" to be separated from the "bread." If the half-model had been plated all it once (on all sides), it would have been impossible to get the center out.

The wax prevented both sides of the horse half from being plated at the same time. To plate the "inside" of the horse half, the "outside" was blocked off with wax—by embedding the horse half, up to the parting line, in wax. The block of wax was about 2" deep and the size of the finished mold (such as, 2.5 feet wide and 18" high). Then, the "inside" (the crater) and the adjoining face of the block of wax were sprayed with silver lacquer. After the silver areas were plated, the wax was melted away. To plate the "outside" of the horse half, the area that had just been plated was covered with wax to prevent it from being plated again (during the second "dip" in the bath). The "outside" (mountain side) of the horse half and its adjoining face of the block of wax were then sprayed with silver lacquer, and were plated in the bath.

Actually, it could be done either way: plating the outside first and the inside, second. Also, both horse halves—left and right—were plated on the same plating sheet. (Think of a cookie sheet with room for two large cookies.) For smaller horses, several horse halves could be plated simultaneously on one plate for the plating bath. All horse halves that were plated on one sheet became part of the same mold.

After plating was finished and the metal horse halves were removed from the center, the resulting copper shell for the "inside" looked like a cup cake mold except that the craters were horse-shaped. The copper shell for the "outside" looked like a cup cake mold turned upside down—a plain with mountains in the shape of horse halves.

Modern Plastics, October 1960, illustrated a "wax master of electroformed injection mold cavities emerging from the copper electroplating bath" at Hartland Plastics. Pictured is a rectangle about 18" high, 2.5' long, and 2" deep with eight impressions of halves of horses from the 5" rider series.

Electroplating took several days. When it was done, "The shell was pried apart, and the Monel metal horse halves removed," Hans Seuthe said. The Monel metal halves would often be damaged in the process, and be of no further use, he said.

Tool Making (Stage 3). According to Hans Seuthe, the tool makers (plating department and machine shop) then machined the copper shell to size (essentially, they smoothed the edges so they would meet), bolted the copper shell halves together, and put them in a steel box of the desired size of the finished mold. In Hartland Plastics' own foundry (the casting room), they filled the box with molten kirksite. (Kirksite is a white die metal alloy that is mostly zinc. Paul Champion said that the thin, copper cavities were typically backed up by 5" of kirksite.) After the kirksite cooled and hardened, they machined the cast block to size and thickness and sawed through the bolts to open it with the copper shells inside. The side with the outside of the horse impression, called the cavity side or female half, received a mounting plate and a hole for the plastic to enter through. The core side or male half had a steel plate, ejector box (to eject the molded, plastic parts), mounting plate, and ejector plate and pins attached to it. Then the mold was ready to be placed in an injection molding machine to mold horse parts.

Hans Seuthe said that, with good luck, it took about 30 days for Hartland Plastics to be ready to make plastic parts, starting from sculpturing, then modifying the sculpture. He said that with the mold-making process used now (in the 1990s) in this country, it would take four to six months or longer. He said that Hartland Plastics' molds were "not as durable," but that they yielded "good detail." The article in *Modern Plastics*, October 1960, said that the electroformed molds cost less than machined steel molds or cast beryllium molds. It concluded:

Hartland's program is of interest in that it shows what can be accomplished when a molder uses creative thinking in organizing his plant and selecting methods of manufacture tailored to the specific problems confronting his business.

Horses From Plastic

The Wonder Material. In the 1967 movie *The Graduate*, the word of advice to the graduate, the field to get into, was "plastics." Plastic is common, but remarkable. It pours, yet retains its shape because it has "memory." From light switches to telephones to car dashboards, it has helped shape modern life.

Hartland's most memorable models were made of high quality plastic. In the 1960s, the 9-11" series horses were always cellulose acetate, a heavy, durable, and high quality thermoplastic known by several brand names, including Tenite. While many plastics are derived from oil, cellulose acetate is made of plant material, typically wood pulp and straw, cotton linters, or vegetable fibers. The smaller horses in the 1960s were usually a less expensive, but relatively durable, styrene, which is technically known as vinyl benzene. Styrene begins as coal tar, natural gas, or petroleum (oil). From the 1970s through the 1990s, both plastics have been used for Hartland horses, also.

The price and availability of model horses can be affected by the world market in oil. When the price of oil goes up, styrene and other thermoplastics become more expensive, but cellulose acetate does, too, because the suppliers of raw materials prefer to supply expensive materials and make less of the lower priced materials available. The lower priced materials then go up in price. When prices go too high, end-product manufacturers tend to cut back on what they consider to be nonessentials (or lower profit items), such as model horses.

Tenite is a brand name for cellulose acetate (and other plastics) made by the Tennessee Eastman division of Eastman Chemical Products Co. (now called Eastman Kodak Company). Edwin Hulbert said that 98% of the acetate plastic Hartland used was Tenite. He said that Hartland Plastics also sometimes used Hercocel, acetate plastic from the Hercules Plastics division of Eastman Chemical or an acetate made by Celanese Corp. The styrene plastic came from Dow Chemical Co., Hulbert said. Both white and colored plastics were used. Colored plastic cost more, but molding a horse in colored plastic reduced the amount of painting the model needed.

Acetate in general was a good plastic for detailed figurines because it was pliable for good matching of molded sections; reproduced details well; and was, according to *Modern Plastics*, "easy on the soft, thermoformed cavities of the mold." It could be buffed and trimmed to eliminate seams, but it was more expensive than styrene. It cost 70 cents per pound compared to 11 cents for styrene in those days, according to Richard Petfalski, who was production manager during much of 1958-1972, and was purchasing agent in his last few years at Hartland Plastics.

Styrene could stress the molds, and Petfalski recalled that, "Styrene was harder to work with." The molding supervisor from 1949-1977, Hans Seuthe, called styrene, "junk for the 'Five and Dime.'" In contrast, "Cellulose acetate won't break," and can take the cold. He said his kids left the acetate horses buried in the sand box all winter (through Wisconsin's below-freezing temperatures), dug them up in spring, and they were O.K. However, acetate is "susceptible to heat." Jerome Delsman, who was night shift plant supervisor from 1949-1972, said the difference between styrene and acetate was like a Chevy vs. a Cadillac. Mrs. Robert McGuire described the unpainted, styrene horses as "made for Woolworth's and sold in bulk."

Petfalski recalled lawsuits due to furniture damaged by plastic hooves sticking and leaving marks during summer hot spells. The plastic would bleed through the paint plasticizers. After that, Hartland put felt pads on the feet of its horses. This was a problem only with the acetate (Tenite) models; the styrene models never damaged furniture, Paul Champion said. From what I've seen, the pads were given only to woodcut horses, which were less likely to be toys and more likely to spend time displayed on good furniture.

(A Hartland horse I got when I was 9 spent most of its time on our upright piano—until I was 11 and became a "collector" by gathering horses from around the house into my bedroom to hold horse shows with them.)

Hartland sent its non-usable scraps of plastic to the town dump. Rejects were "short shots" and painted models that did not meet quality standards. ("Short shots" were molded pieces that were incomplete because the plastic had not been under enough pressure to completely fill the extremities of the mold cavity.) The rejects were often buried, but sometimes employees took them home, which was allowed. Rejects that were clean and not painted were ground up, and their plastic was re-used. The reground plastic was a little darker in color.

According to Jerome Delsman, in about 1967, some 10,000 painted and packaged baseball player statues—six or seven truckloads—were taken to the dump because Hartland Plastics "couldn't give them away." No molds were ever taken to the dump, though.

Employees were able to purchase finished and packaged models at the same wholesale price offered to stores, which was just under half of the retail price. In the late 1960s, bags-full of halves of horses, mostly 5" series

styrene horses, were also sold to employees. Collector Barbara Zoulek got some that way because her father knew a Hartland employee. She said the bags included a lot of 5" Arabian mares which she glued together, painted, and sold to friends.

Molding and Assembly. The plastic, which arrived in beads one-eighth of an inch in diameter, was heated until it was softened to the consistency of toothpaste, and then rammed into the mold at 100 tons pressure. (Hans Seuthe said that the molding temperature of cellulose acetate ranged from 280-380 degrees F; for styrene, it was 380-480 degrees F.) The automated molding press made one piece about every 45 seconds to a minute for things like horses, Edwin Hulbert said. For items requiring less cooling time, pieces could be made every 10 to 20 seconds. After cooling, sections of acetate (Tenite) plastic were glued together with acetone (a solvent that bonds acetate plastic and is also the major ingredient in finger nail polish remover). The solvent used to cement styrene was foluene. Then, excess plastic was removed, and seams were smoothed. (That was called trimming and buffing.) Then, models were sent to the paint department, except for models that were sold unpainted.

Painting. Hartland Plastics painted models at large, individual spray booths or in line booths. One color was sprayed on at each booth, using the fitted, copper electroformed masks. Pencil spray guns were used for antiquing, Paul Champion said. The paint dried in 15-20 seconds, and the models moved by conveyer belt to the next stop. Gloss coats (of clear lacquer) were added to some models; the acetate paint, itself, was semi-glossy. According to *Modern Plastics*, there were 70 painting stations. At the end of the painting line, the model was packed and prepared for shipping. The paint came in five gallon pails, but the clear lacquer came in 55 gallon drums. The colors of the paint and the plastic could vary between batches. Hartland Plastics used eight colors of fast-drying paint from Bee Chemical Co., Chicago, Illinois, and Wolverine Finishes Corp., Grand Rapids, Michigan.

Colored Plastic. About one-half of the 1950s-1960s Hartland horses were made of colored plastic (black, golden-yellow, pale taupe, reddish brown, etc.) instead of white plastic, as a painting short cut. Typically, they would paint the points and details (mane, tail, lower legs, hooves, eyes, etc.) and leave the body the molded color. Many 7" and smaller Hartland animals were also issued entirely unpainted in styrene plastic of various colors.

Manes and Tails. On breed-series horses, most had tails with each half of the tail molded as part of the left or right half of the horse. Those tails have a seam down the center of the tail, as viewed from the top or back of the tail. I call them two-sided tails. The two-sided tails are found on the 11" series Arabian, Quarter Horse and Saddlebred, and on all of the 9" breed-series horses except three. On the 9" Arabian Stallion and Three-Gaiter (woodcut), the entire tail is molded as part of the right side of the horse. The 9" Five-Gaiter has its tail entirely on the left side. The model-making department, including the mold engineer and sculptor Alvar Bäckstrand, decided how to divide the sculpture into molded sections. In some cases, Bäckstrand would alter the original model he or Roger Williams had made so that the model could be molded more easily.

On Hartland horses, most manes appear on both sides of the neck or completely on the right side. Customarily, for English riding, the mane is brushed to the right to keep it from getting tangled in the reins when mounting or dismounting. For western riding, the mane is worn on the left since a western rider might grasp some of the mane when mounting, and western reins as you mount and less likely to get tangled with the mane.

The 9" Tennessee Walkers had two different types of manes. When I asked Hartland officials how that was done, I was told that they kept a metal version of each model that could be duplicated, and that molds could be made from molds. In other words, they didn't have to start from scratch with another clay horse with the new mane. Sculptor Alvar Bäckstrand said, "Small changes could be made in molds by adding copper and grinding off."

Woodcut Horses. The look of carved wood was created artificially. Sculptor Alvar Bäckstrand said, "I cut the wood-carved look into the metal model that was going to be used for the plating of the mold. I used an air tool to cut the wood-carved look, and the wood grain. There were no horses ever made of wood [by Hartland Plastics sculptors] for the mold model."

Palomino and Pearl White. Since the translucent, palomino color would not flow evenly unless heated, it was reportedly saved for the larger horses, and an opaque, orange-yellow was used on the Tinymite Morgan. Edwin Hulbert said that the translucent pigments had to be heated to 150-175 degrees. The paint for the pearl white horses was reportedly $35 for 16 oz. of paste. The color was used sparingly: on a 9" Arabian Mare, an 11" Arabian Stallion, and on some religious statues.

Marketing Decisions. Robert McGuire and Paul Champion suggested and approved the colors on the models. They would do sales forecasts, put test colors on the market, select the final colors, write the sales literature, and arrange for the catalog photography. The sample models were spray-painted, using masking tape to block off the color areas. Horse colors were taken from photos which were given to the paint department to try to match.

Talented People

Hartland Plastics had assembled "a lot of talented people. Talents that don't exist anymore," said Hans Seuthe, molding department foreman from 1949-1977. First on the list of talent were the two sculptors. Hans Seuthe recalled the excellent religious figurines and animals by Roger Williams, and called Alvar Bäckstrand a genius in his knack for faces, seen in the ballplayers and riders. He said Alvar did unbelievable things with wood. The wood carvings were done for fun, in spare moments at the plant; figures to be molded were almost always done in clay.

Richard Rohde said that his father, Armin, was proud to assist Alvar Bäckstrand in the model department. The older Rohde's extensive experience in making dental castings, inlays in metal, was valuable for the model work with plaster and metal prior to plating. Armin Rohde was a model maker at Hartland Plastics from 1955 or 1956 until he retired in 1974. Richard Rohde also worked at Hartland Plastics for a time.

Peter Radix, who came from The Netherlands, cut out the paint masks that defined the painted details so neatly. The paint masks were made of copper. Leonard Noll also did mask design work. Peter's son, Hein Radix, supervised the plating department. Hein's sister, Kitty Radix, worked in the assembly and painting department, and packed models into individual cartons. She was remarkably fast at packing and "regularly did as much as two normal people could do."

As first shift supervisor for the molding department, Hans Seuthe critiqued the molds with the aim of making them more runable. He was involved with tooling, sampling with the molding machines, and getting parts running. It was important to maintain the molds, and it took skill to do so, he said. Until the late 1980s, he had saved the production records, records of what was made when, and in what quantities.

"Good tool and die men worked under Ed Hulbert," said Richard Petfalski, who expedited production between 1958-1972. That included John Timmerman, Bill Lewis, Charles ("Bud") Delsman, and Richard Schesler, foreman of the tooling department. Jerome Delsman was supervisor of the molding department for second and third shifts, from 3 p.m.-7 a.m.; he recalled his dates at Hartland Plastics as January 3, 1949-April 10, 1972. Edgar Schmidt was a molding foreman and made the assembly fixtures for the horses. Carl Postulart was a machine setup man and then second-shift foreman. He was from The Netherlands and was married to Kitty Radix. John Scullin was the draftsman engineer who made drawings for the mold parts, including for horses and riders.

After the parts were molded, the finishing department glued them together and did the painting, quality inspection, and packaging. John Nicholas was the finishing department supervisor during the religious statue and Horse/Rider era, Edwin Hulbert said, and throughout the model horse era. About 1969, John Nicholas moved to the office as production manager. George Gerbenskey did all the paint mixing before automation, when it was a skill, and painted the first figures for selling purposes, Alvar Bäckstrand said. Beaman Meatcham also mixed paints. Anita Marquardt (Mrs. Jerome Delsman) made sure that models were properly glued together. She worked out the procedure for each new product and trained the assemblers on all three shifts. Ann Thurloff was the quality control inspector. Edwin Hulbert said that she efficiently kept track of things in the molding and painting departments, such as making sure that the parts coming off the molding machine were of sufficient quality to work in the paint masks. She brought the painted models to Paul Champion (or Robert McGuire) for approval of colors, and she kept the models within quality limits.

Ann Thurloff was also was in charge of personnel in the large painting department. Hartland Plastics was a popular workplace for young women who worked in the painting department or in the office for a few years between school and marriage. Married women worked there also. Janet Gerbenskey, sister-in-law of George Gerbenskey, recalled that she spray-painted models and other items for 20 years, from July 1955-September 1975. The office staff included Eldora (Ellie) Perry, Paul Champion's secretary; Nancy Luko, Robert McGuire's secretary; and Evelyn (Evie) Wendt, who was office manager and, as Paul Champion said, "mother to everybody." Joseph Mosciski was bookkeeper. Shirley Lugner, Joanne Roets, Janice Adams, Frances Hoffman, and Carol Palmer also worked in the office at various times in the 1950s or 1960s. With the fast turnover in the young painting staff, it's probable that 1,000 people worked at Hartland Plastics during the company's 40 years.

Conclusion. Hartland Plastics' two sculptors did memorable work. Horses are particularly difficult to sculpture, but Roger Williams captured breed characteristics and the movement of horses with great flair. Alvar Bäckstrand had a special talent for sculpturing faces that were good likenesses. An able group of craftsmen—including Hein and Peter Radix, Armin Rohde, Leonard Noll, and Hans Seuthe—and attentive supporting staff enabled the transition from the artist's original to shelves or toy boxes of Hartland admirers all over America.

Collecting model horses doesn't consist just of shopping for them and then admiring them on a shelf. We might name them, assign them parents, decide what breed or gender they are (if those things are in question), and generally enjoy them in creative or imaginative ways. Since 1970, there have been clubs, registries, and news journals by and for model horse collectors. Since the mid-1990s, collectors have also been linked by model horse discussion, reference, and sales sites on the Internet.

Model Horse Shows

A major activity of the hobby has always been model horse shows, of which there are two types: photo shows, in which photos are mailed to judges, and in-person judging meets, which are called "live shows" or "model horse expos." The main purpose of these expos is not to buy or sell models, but to pretend that the event is a real horse show with real horses being judged. Of course, this requires many accommodations for the differences between models and real horses.

Entrants at expos are typically 10 to 60 years old. They are competing for ribbons or trophies that may be larger than some of the models. Model shows are not static exhibits; 50 to 100 or more different competitions (called "classes") may take place in one day with new classes being called to the judging tables every 10 or 15 minutes.

The judged events at live shows typically include collector classes, halter classes, and performance classes. Collector classes are open only to original finish models, which are judged on their condition, rarity, and age. These classes chiefly consider the models as collector items.

Halter and performance classes, on the other hand, are judged as miniature, still-life versions of real horse show competitions. In halter classes the models are judged on how closely their body build conforms to ideal standards for their breed, as real horses are. Separate halter classes are held for original finish models and customized (repainted or remade) models, and within those two categories, there are many separate classes for the different breeds being represented.

Remade models are those that have been repositioned by heating and then bending or moving parts such as the head and legs. This should not be attempted without instruction because of the danger of burns or the model bloating. Hair-like manes and tails are sometimes added for more realism. In halter classes for remade or repainted models, neatness and accuracy of the alterations are taken into account.

For judging halter classes, the models are simply placed on the judging table. (Classes for 4-H members may require a halter or bridle appropriate to the breed, but for most halter classes, the model itself is the only item needed.) For performance classes, however, the models are always outfitted in miniature tack—saddles, bridles, and other equipment appropriate to the class. The models are equipped with English saddles and bridles for English classes and western tack for western classes. Harness classes sometimes require a vehicle in addition to the harness. Riders, drivers, or handlers are not required, but appropriately attired dolls are occasionally seen. Costume classes inspire many imaginative entries. Since hardly any tack is made commercially, collectors make their own or buy from some very talented tack-makers within the hobby. The winning entries are those with the most realistic, appropriate, and correctly-scaled tack and that best depict the action called for in the class.

Of course, models can't move, but their pose should depict a frozen instant of the action that would be seen in that particular class in a real horse show. For instance, in a park class, the winners will be elegant models—Saddlebreds, Arabians, and Morgans, for the most part—in high stepping poses or alert, standing poses. In a western pleasure class, on the other hand, the best entries are calm-looking models standing quietly or posed in a low-action jog or relaxed lope.

Some performance classes also require props in addition to the tack on the horse. In jumping classes, for instance, each model must be accompanied by a fence or other obstacle supplied by the model's owner. The obstacle also is considered in the judging. It should be realistic and correctly scaled to the model's size. The placement of the model in relation to the obstacle is also important. The scene should look as though the model could approach and clear the obstacle. Many exhibitors make their own jumps out of balsa and put miniature, potted plants in front of them for a decorative touch that is sometimes seen at real horse shows.

Model showing gives collectors a chance to demonstrate their understanding of real horse show requirements and procedures by depicting it with models. Many of them study the real horse judging standards of the American Horse Shows Association and numerous horse breed associations and then apply them to models. For each class they plan to enter, the exhibitors select a suitable breed and pose of model from their collection, outfit it authentically—usually minus the rider—and provide any required props.

Shows are held by individuals, model clubs and 4-H clubs, and at the annual BreyerFest. The model hobby now has an organization called the North American Model Horse Shows Association, the purpose of which is to promote and standardize model showing. NAMHSA holds a national show each year, the North American Nationals, for which models qualify by winning a first or second place at a member show. Model horse shows are also held in England, France, Germany, Sweden, Australia, and other horse-loving countries.

A Creative Hobby

I read somewhere that at any given time, 20% of girls 10-14 years old and 10% of boys ages 8-12 in the United States and Europe are horse crazy. Reflecting back to school days, the statistic seems to hold true. For many, horses are a lifetime interest that can take many forms of expression, with model horses being one of them. Going far beyond "playing with toys," the hobby can be very sophisticated. Planning, organizing, and managing a "show stable" of models—or even a collection just for fun—can be a complex operation. The model hobby can be an outlet for craftsmanship and creativity whether the purpose is to enter shows or not, and it typically invites research into the horse world.

As a substitute for horse ownership or riding lessons, model collecting is generally less expensive and certainly more convenient. However, models are not necessarily a substitute for real horses or a preparatory phase to be abandoned later. There are horse owners—even breeders—who enjoy model collections, too. A fair portion of model collectors have never been heavily involved with real horses. Models are art, and they have their advantages: less maintenance, and you don't need a protective helmet. Models won't dump you in the dust although they may need dusting. The two hobbies—horses and model horses—go together well.

Model Horse Hobby Resources

The Hobby Horse News: Tina Ferro, The Hobby Horse News, 14 Garraux Street, Greenville, SC 29609; THHN2000@webtv.net. Bi-monthly news, features, and ads; for several years, the annual Special Edition has listed model sellers, tack makers, customizers, clubs, shows, books, newsletters, etc.

The Online Model Horse Resource Guide — www.starrfyre.com/rds/resource.htm — copyright 1999-2000 by Beth Gustas. Lists discussion, reference, and public sales web sites; online model dealers; and many off-line hobby resources such as clubs, books, photo and live shows, etc.

Also, see the Bibliography.

Note: This chapter is an updated version of my article in The Milwaukee Journal, *July 14, 1983.*

Bibliography

Articles in Magazines/Newspapers/Newsletters

Chip Chats (National Wood Carvers Association), July-August and Sept.-Oct. 1981.

English, Tina. Hartland news and notes in *High Stepper's Review*, July-Aug. and Nov.-Dec. 1990.

Fitch, Gail. Articles (too numerous to mention) on Hartland horses, dogs, and company history in 18 issues (194 pages) of *Hartland Market*, Dec. 1994-May 1996. For back issues, email: gfitch@execpc.com or write: G. Fitch, 1733 N. Cambridge Ave., #109, Milwaukee, WI 53202.

Fitch, Gail. "Collecting Hartlands: A Storied Breed," *The Model Horse Gazette*, March/April 1996.

Fitch, Gail. Hartland column (11 articles) in *TRR Pony Express*, April 1994-March 1996.

Fitch, Gail. "Horses without the barn essentials," *The Milwaukee Journal*, July 14, 1983.

Fitch, Gail. "Riding High with Hartland," *Toy Collector and Price Guide*, October 1995.

Fitch, Gail. "Why Hasn't Someone Bought Hartland Plastics?," *(White's Guide to) Collecting Figures*, May 1996.

Groeber, Paola. "Let's Talk Hartlands," *Model Horse Gazette*, July/August 1985, November/December 1985, and July/August 1987 (which corrected the quantities of models listed in her Christmas 1986 brochure).

Groeber, Paola. "What's Happenin' with Hartlands." *American Model Horse Collector's Digest*, July/August 1987.

"Holy Image on the Dashboard," *Life* magazine, April 20, 1959.

National Sculpture Review, Summer 1972, Summer 1973, Spring 1978, and Second Quarter 1988. Also, *Sculpture Review*, Spring 1986.

"New Developments [Hartland Plastics]," *Modern Plastics*, October 1960.

Novick, David. "Plastics in Wisconsin: The Squeeze is On," *Investor, Wisconsin's Business Magazine*, May, 1974.

Tomezik, Sandy. "Hartlands: A Fantasy in Plastic," in Shari Struzan's *The Hobby Horse*, October 1980.

Van Etten, Rick. "Hartland *Isn't* Just for Horses Anymore," *Hartland Market*, October 1995 (Hartland dogs).

Vintage articles in *The Hermann Advertiser-Courier*, Hermann, Missouri; *The Lake Country Reporter*, Hartland, Wis., *The Milwaukee Journal*; *Oconomowoc Enterprise*, and *Waukesha* (Wis.) *Freeman*, particularly an article on Alvar Bäckstrand in the February 6, 1971, *Waukesha Freeman*.

Books

American Kennel Club. *The Complete Dog Book*. New York: Howell Book House, 18th edition, 1992.

Bikales, Norbert M., editor. *Molding of Plastics*. New York: Wiley-Interscience, 1971.

Birren, Faber. *Principles of Color*. West Chester, Pennsylvania: Schiffer Publishing Ltd., 1987.

Brown, Paul. *Draw Horses*. New York: Charles Scribner's Sons, 1949 (equine gaits).

Classified Directory of Wisconsin Manufacturers. Milwaukee, Wis: Wisconsin Manufacturers' Association, 1948-1978 annual directories.

Edwards, Gladys Brown. *Anatomy and Conformation of the Horse*. Croton-On-Hudson, New York: Dreenan Press Ltd., 1973.

Fitch, Gail. *Hartland Horsemen*. Atglen, Pennsylvania: Schiffer Publishing Ltd., 1999.

Fitch, Gail. *Hartland Horses and Riders*. Milwaukee, Wis., and Santa Barbara, California: six self-published editions (20 printings), 1983-1995.

Fitch, Gail. *Hartland Models, Vol.1/Hartland Horsemen & Gunfighters/* Milwaukee, Wisconsin: self-published, 1998.

Fitch, Gail. *Horse Colors and Gaits*. Milwaukee, Wisconsin: self-published; three regular editions, 1985-1990, and two abridged editions, 1994 and 1995, *Gift Horses from Japan*.

Harris, Julie and Frederick. *Resin Registry*. Tampa, Florida: self-published, 1997 with 1998 and 1999 updates (reference on resin-cast model horses). Julie A. Harris, 4218 Interlake Dr., Tampa, FL 33624; julhar@mindspring.com.

Hetzler, Bruce, Ph.D. *Bev Taylor's Town House Magic*. Appleton, Wis.: self-published. Bruce Hetzler, 43 Bellaire Ct., Appleton, Wis. 54911.

Holland, Thomas W., editor. *More Boys' Toys of the Fifties & Sixties: Toy Pages from the Great Montgomery Ward Christmas Catalogs, 1950-1969*. Sherman Oaks, California: The Windmill Group, Inc., 1998 (includes Hartland copies).

Katz, Sylvia. *Plastics/Designs and Materials*. London: Studio Vista, 1978.

Kays, D. J. *The Horse: judging, breeding, feeding, management, selling*. New York: Reinhart, 1953.

Kenny, Colleen. List of Breyer Models. Deerfield, Illinois: self-published, 1976.

Made In Japan [Horse] Club. *China [Horse] Reference Book*. Chapel Hill, North Carolina: two self-published volumes, 1996 and 1998 (includes metal copies of Hartlands). Barri Mayse, PO Box 573, Dayton, Tennessee 37321; Amaysen3@aol.com.

Palley, Reese. *The Porcelain Art of Edward Marshall Boehm*. New York: H.N. Abrams, 1976.

Smith, Charles W. *Auctions: The Social Construction of Value*. New York: The Free Press; London: Collier Macmillan Publishers, 1989.

Sponenberg, D. Phillip, and Bonnie V. Beaver. *Horse Color*. College Station, Texas: Texas A & M University Press, 1983.

Sponenberg, D. Phillip. *Equine Color Genetics*. Ames, Iowa: Iowa Sate University Press, 1996.

Turner, Diane E. *Understanding Your Horse's Lameness*. New York: Arco Publishing, Inc., 1980.

Waring, George H. *Horse Behavior*. Ridge, New Jersey: Noyes Publications, 1983.

Wynmalen, Henry. *The Horse in Action*. New York: Barnes, 1954.

Young, Jean and Jim. *The Garage Sale Manual: alternate economies for the people*. New York: Praeger Publishers, 1973.

Government Records and Civic and Non-Profit Organization Web Sites/Publications

Durant, Oklahoma: www.durant.org

Hartland, Wisconsin: www.hartland-wi.org

Hartland: A Thematic History. Waukesha, Wis: The Southeastern Wisconsin Regional Planning Commission, 1985.

Hermann, Missouri: www.hermannmo.com

National Sculpture Society—www.nationalsculpture.org—1177 Avenue of the Americas, New York, N.Y. 10036; (212) 764-5645.

Register of Deeds, Waukesha County, Wisconsin.

U.S. Patent and Trademark records.

Sales Literature/Company Catalogs

Anheuser Busch gift catalogs. (Call 1-800-325-9665.)

Cascade Models' ad in Tina English's *High Stepper's Review*, September/October 1991 (explains Christmas 1985 models).

eBay, an Internet auction site: www.ebay.com (illustrated auctions viewed in 1998-2000).

Hartland and other model horse manufacturers'/distributors' catalogs and advertisements, 1950s-present.

Hartland Model Equestrian—www.execpc.com/~gfitch/Index.htm—© 2000 by Gail Fitch.

Mission Supply House history (www.collect-us.com)

Nylint Corporati